Introduction to Real Analysis

The Prindle, Weber & Schmidt Series in Mathematics

Swokowski, *Fundamentals of College Algebra, Seventh Edition*
Swokowski, *Fundamentals of Trigonometry, Seventh Edition*
Swokowski, *Precalculus: Functions and Graphs, Sixth Edition*
Tan, *Applied Calculus, Second Edition*
Tan, *Applied Finite Mathematics, Third Edition*
Tan, *Calculus for the Managerial, Life, and Social Sciences, Second Edition*
Tan, *College Mathematics, Second Edition*
Trim, *Applied Partial Differential Equations*
Venit and Bishop, *Elementary Linear Algebra, Third Edition*
Venit and Bishop, *Elementary Linear Algebra, Alternate Second Edition*
Willard, *Calculus and its Applications, Second Edition*
Wood and Capell, *Arithmetic*
Wood, Capell, and Hall, *Developmental Mathematics, Fourth Edition*
Wood and Capell, *Intermediate Algebra*
Zill, *A First Course in Differential Equations with Applications, Fourth Edition*
Zill, *Calculus with Analytic Geometry, Second Edition*
Zill, *Differential Equations with Boundary-Value Problems, Second Edition*

The Prindle, Weber & Schmidt Series in Advanced Mathematics

Brabenec, *Introduction to Real Analysis*
Eves, *Foundations and Fundamental Concepts of Mathematics, Third Edition*
Keisler, *Elementary Calculus: An Infinitesimal Approach, Second Edition*
Kirkwood, *An Introduction to Real Analysis*

This book is dedicated to my wife Bonnie. Without her love and support, this book would not have been written.

Introduction to Real Analysis

ROBERT L. BRABENEC
WHEATON COLLEGE

PWS-KENT Publishing Company
Boston

PWS–KENT
Publishing Company

20 Park Plaza
Boston, Massachusetts 02116

PWS-KENT Publishing Company is a division of Wadsworth, Inc.

Library of Congress Cataloging-in-Publication Data

Brabenec, Robert L.
    Introduction to real analysis / Robert L. Brabenec.
       p. cm.
    Includes bibliographical references.
    ISBN 0-534-92162-0
    1. Functions of real variables.   2. Mathematical analysis.
    I. Title.
QA331.5.B73 1990   89-29582
515′.8—dc20   CIP

Printed in the United States of America.

90 91 92 93 94—10 9 8 7 6 5 4 3 2 1

Sponsoring Editor   Steve Quigley
Production Editor   Helen Walden
Text and Art Preparation   Robert L. Brabenec
Cover Design   Designworks
Manufacturing Coordinator   Ellen Glisker
Printing and Binding   Maple-Vail Book Mfr. Group

To the Teacher

This book is intended for a first course in analysis, following two or three semesters of calculus, and is suitable for a variety of course structures. Chapters 2–4 and 8 could be used for a one-quarter course on the nature of proof in analysis. There are several possibilities for a one semester course. One would be to cover Chapters 1–6, with an emphasis on sequences, continuity, derivatives, and integrals. Another would be to cover Chapters 1–4 and 7–9, with an emphasis on convergence, assuming the basic facts about derivatives and integrals that are covered in calculus. A third possibility is to cover most of the book, by deleting or providing minimum coverage of such sections as 12, 13, 22, 25, 26, 28, and 36. The introductory material in Chapter 1 and the historical introductions in Sections 3, 14, 18, and 23 can be omitted from any course, or left for the student to read. Most of the book could be covered in a two-quarter sequence, and the book is suitable for a year course, if you proceed at a more leisurely pace, or supplement the book in a few places.

It is my goal to begin the study of analysis as soon as possible, which I do in Section 4. The appendices are available, if needed, for terminology about sets, the real numbers, and logic and proof. In Section 5, I develop many basic results about sets, which are used repeatedly throughout the rest of the book. In Section 9, I introduce the idea of infinite series in the setting of the study of sequences. If desired, you could go on to cover Chapter 8 on infinite series immediately, before returning to begin the study of functions of a real variable in Chapter 3.

I do not believe it is necessary to prove every theorem in a first course at this level. I have included only the statements of several theorems, such as Theorems 22.5, 26.9, 30.6, and 31.7, that have fairly complicated proofs and are such that the proofs are not especially enlightening for the student. I do include a reference where such proofs may be found. In addition, you may choose to discuss only representative proofs in such places as Sections 5, 11, 13, 22, 25, and 26, where there are many theorems with similar proof forms.

It is often helpful to a student in analysis if you include some classroom activities that balance the usual straight lecture approach. One possibility is to have the students work at the board on a problem or proof, either individually or as a class. In this way, you can better see the way that their mind works. Another possibility is to spend time discussing a more challenging example or exercise, or showing alternative ways to prove a given result. The book is written using an uncluttered, reader-friendly approach, so that the student is able to understand the basic ideas in each section. This should free you to provide some alternate activities as described above in order to improve their understanding.

I chose to number the sections consecutively throughout the book, using the chapter headings as subdivisions. All results that call for a proof (theorem, lemma, or corollary) are numbered consecutively within a section, using the section number as part of the designation. For example, Theorem 24.3 is the third result requiring a proof in Section 24. Definitions, examples and exercises are numbered separately and consecutively within a given section by a single number, using the section number only when referring to such an entity in a later section. For example, Definition 2 denotes the second definition in Section 4, but I use Definition 4.2 when referring to this definition in a later section.

I hope that this book fulfills your expectations and works well in your class setting. I would appreciate receiving your suggestions for ways in which the book could be made more useful to you.

To the Student

This book will introduce you to a study of the theory of calculus. While you have already studied two or three semesters of calculus, the emphasis there was on an intuitive understanding of the material. The concept of a limit is the distinguishing characteristic of calculus, and a careful epsilon treatment of the theoretical implications of the limit concept constitutes an introduction to real analysis.

You will notice that several sections are devoted to historical background, and additional historical comments are interspersed throughout the rest of the book. I have found that the history of mathematics is not only interesting, but that it helps me to understand the theoretical results better. I hope that your experience will be the same. The bibliography contains suggestions for further reading on your part to supplement my comments.

Chapter 1 is intended to begin the transition from an intuitive calculus to a more rigorous analysis. Typical problems from calculus are given to help you review the material that you studied a year or more previously, along with comments about the main theorems and the historical development of calculus. You will not master this material at all after a first reading. I recommend that you refer back to this chapter frequently throughout the course.

Chapter 2 is perhaps the heart of the course. Most of this material will be new to you, since you are just beginning your study of careful proofs in analysis in the context of sets and of sequences of real numbers. You should also refer to the three appendices as needed to provide background information. You will undoubtedly be uncomfortable during your study of Chapter 2, always feeling that you do not have any kind of mastery of the material. This is normal—as you continue your study in the succeeding chapters, the ideas in Chapter 2 will become much clearer.

Chapters 3 and 4 extend the idea of epsilon proofs from the topic of sequences to the setting of functions of a real variable, including the important idea of continuity. As you work through the problems and theorems in these two chapters, you will see many analogies to the comparable results in Chapter 2, and also why the topic of topology is useful as a means to unify the various forms of the limit definition and theorem.

The most familiar material will be that on the derivative and integral in Chapters 5 and 6, with such results as the Mean-Value Theorem,

L'Hospital's Rule, and the Fundamental Theorem of Calculus. There will also be new material, such as Taylor's Theorem with Remainder, and the careful proofs of many properties of the definite integral.

Chapters 7–9 deal with the idea of convergence that was begun in Section 9. Much of the material on infinite series in Chapter 8 is covered in a calculus course, but most calculus students have considerable difficulty with this topic, and profit from a second presentation here. There are many interesting applications in these chapters, which I am sure you will enjoy very much.

The exercises at the end of each section provide an excellent means for learning the material. They consist of both calculation-type problems and proofs of results, so that you can develop your skill in both areas. There are pages at the end of the book consisting of hints and answers for selected exercises. You should attempt the exercises on your own however, before resorting to these helps.

I suggest that you read through each section before your teacher lectures on it, noting especially the defined concepts, the statements of the theorems, and the kind of examples that are presented. You will profit much more from the lecture if you have prepared yourself in this way. After the lecture, you should reread the section, and spend more time on the details of the proofs and of the examples. You will probably read through some proofs several times. Most proofs contain one or two key ideas, and it is important that you seek to identify these, especially for the major results, which are called theorems. The term *lemma* refers to a less important result, which is used to help prove a theorem. The term *corollary* refers to a result that follows quickly as a consequence of a theorem. The symbol ■ is used to mark the end of a proof for a lemma, theorem, or corollary. The symbol □ is used to indicate the completion of an example.

I conclude with the encouragement that you are not assumed to understand everything that you read in this book (or other mathematical books), even if the author says that it is obvious. Mathematical learning is a gradual process and the same ground must often be tread upon many times before understanding occurs. A first theory course in analysis is one of the most challenging courses that a mathematics major will take, but if there are many difficulties ahead of you in this study, there are also many rewards for a diligent student. I wish you well as you begin your work.

Acknowledgments

I would like to express my appreciation to many people who have helped me with the writing of this book. R. Creighton Buck from the University of Wisconsin at Madison, my Wheaton colleague James Mann, and A. Wayne Roberts of Macalester College all read early drafts of some of the manuscript, and helped to start me in the right direction. Dale Varberg of Hamline University has offered wise counsel to me at several points of the writing. Ted Vessey of St. Olaf College read most of the manuscript and made many good suggestions for improvement and revision.

Several people read selected parts of the book in a more finished form, and helped to significantly improve the clarity and accuracy of the writing. These include William Ballard (University of Montana), James Carlson (Creighton University), John Duncan (University of Arkansas), Charles Hampton (College of Wooster), Charles Himmelberg (University of Kansas), Richard Hodel (Duke University), Russell Howell (Westmont College), Calvin Jongsma (Dordt College), L.R. King (Davidson College), William E. Mastrocola (Colgate University), John Mathews (California State University at Fullerton), Gordon Melrose (Old Dominion University), Roger Nelsen (Lewis and Clark College), Roger Nelson (Ball State University), W.F. Pfeffer (University of California at Davis), Richard Stout (Gordon College), Dale Varberg (Hamline University), and Paul Zwier (Calvin College).

Since I typeset the book myself, I needed frequent help with the computer system to which I entrusted my manuscript. John Hayward of the Wheaton College computer science faculty was always available to provide the help that I needed. I am also grateful to my analysis students over the past twenty five years, who shaped the form of the book through their input in the classroom and in my office.

It has been a pleasure to work with the staff at PWS-KENT Publishing Company. They have always given prompt and helpful responses to my questions. Special thanks go to the mathematics editor, Steve Quigley, who has guided the entire project in a most professional way, and yet given me the freedom to make many of my own decisions.

As any author knows, a writing project requires a commitment that puts demands on a family life. I am glad for an understanding family and for the encouragement of my wife, my children, and my parents during the past three years that I have worked on this manuscript.

Contents

1

A Review of Calculus
and An Overview of Analysis

SECTION 1 AN INTRODUCTION TO THE STUDY OF ANALYSIS

Analysis is the study of the limit concept and its ramifications. These include the following four important terms that are defined by means of a limit—the *continuity* of a function, the *derivative* of a function, the *definite integral* of a function defined on a closed interval, and the *convergence* of infinite sequences and series. The major theorems of analysis are usually statements about the relationships between pairs of the above concepts. As one example, the Fundamental Theorem of Calculus is a statement about the inverse nature of the processes of differentiation and integration.

This perspective determines the structure of the first part of this book. First, one of the major concepts is introduced and illustrated by many examples. Then the definition is reworked so that the basic properties of the concept can be established. Finally, all these results are used to formulate the theorems of analysis. The results in one unit serve as a basis for those in the next unit. For example, the concept of the derivative is introduced after the unit on continuous functions, where we prove such results as the Extreme Value Theorem and the boundedness of continuous functions on closed intervals. After defining the derivative, we illustrate it by obtaining the derivatives of such specific functions as x^3, $\sin x$, $\sqrt{x}$, and 2^x. Next, the definition of the derivative is used to obtain the Product Rule, the Quotient Rule, the Chain Rule, and the fact that a function with a derivative at a point must also be continuous at that point. Finally, these derivative results are combined with the theorems from the chapter on continuity to derive new important results such as the Mean Value Theorem, L'Hospital's Rule and Taylor's Theorem with Remainder.

We now list several main theorems of calculus. Most will be familiar, at least by name, to the student who has finished a course in calculus. The main purpose of this course is to use careful proofs to provide a rigorous development of these results.

1

Result 1 (The Limit Theorem)
The limit of the sum (or product, or quotient) of two functions is equal to the sum (or product, or quotient) of the limits of two functions.

Result 2 (Differentiability implies continuity)
If $f'(c)$ exists, then $f(x)$ is continuous at c. The converse of this result is not true—consider $f(x) = |x|$ at $c = 0$.

Result 3 (Differentiation and integration are inverse processes)

$$\int_a^b \frac{d}{dx} F(x)\, dx \;=\; F(b) - F(a) \;=\; F(x)\Big|_a^b$$

$$\frac{d}{dx} \int_a^x f(t)\, dt \;=\; f(x)$$

The two formulas in Result 3 constitute what is called the Fundamental Theorem of Calculus. The hypotheses are omitted here, so that you can more clearly see how the integral and derivative signs "cancel each other" in either order. The theorems are carefully stated and proved in Section 27.

Result 4 (Continuity implies integrability)
If $f(x)$ is continuous on $[a, b]$, then $\int_a^b f(x)\, dx$ exists.

Result 5 (The integral is a continuous function)
If $f(t)$ is integrable on $[a, b]$, then $F(x) = \int_a^x f(t)\, dt$ is continuous for all $x \in [a, b]$.

As we attempt to formulate proofs of the theorems stated above, we see the need for several results that are essential to achieve these proofs. These results can be considered as helping theorems—they are important because they contribute to the proofs of the main theorems.

Result 6 (Mean Value Theorem for derivatives)
If $f'(x)$ exists on $[a, b]$, then there is a value c in (a, b) so that

$$f'(c) \;=\; \frac{f(b) - f(a)}{b - a}$$

Result 7 (Mean Value Theorem for integrals)
If $f(x)$ is continuous on $[a, b]$, then there is a value $c \in (a, b)$ so that

$$\int_a^b f(x)\, dx = f(c) \cdot (b - a)$$

Result 8 (Extreme and Intermediate Value Theorems)
Let $f(x)$ be continuous on $[a, b]$ with $m = \text{glb}\, \{f(x) : a \leq x \leq b\}$ and $M = \text{lub}\, \{f(x) : a \leq x \leq b\}$. Then for any value y so that $m \leq y \leq M$, there is a value $c \in [a, b]$ so that $f(c) = y$.

The above statement is called the Extreme Value Theorem when $y = m$ or M, and is called the Intermediate Value Theorem when $m < y < M$. Definitions for the terms glb and lub are given in Section 5.

The above theorems are probably familiar to most students of calculus. There are other theorems that you will meet for the first time in this course. Some, such as the Weierstrass M-Test, are concerned with uniform continuity and uniform convergence. Another new topic deals with infinite sequences and series of functions, and includes theorems such as the interchange of limit theorems in Sections 30 and 34.

In order to establish the above results, we find it necessary to examine the axioms on which analysis is based. These include the algebraic concept of the real numbers as a complete ordered field and a result called the least upper bound axiom. These axioms can be used to prove some fundamental results, two of which are the Bolzano-Weierstrass Theorem and the Heine-Borel Theorem. Also, the theory of analysis motivated, at least in part, the development of the important field of topology with its terminology of neighborhood, open and closed sets, and compactness. These ideas in topology serve as a generalization of the analysis concepts in the setting of the real numbers, the Euclidean plane, or n-space.

The study of analysis involves many topics other than mathematical ones. Philosophical ideas are close at hand as we consider the nature of the infinite through the limit concept. The logical validity of a proof by contradiction must be considered at several crucial points. The idea of creativity is also present as we try to understand how these theorems of analysis were first discovered, and as we look for patterns in our work. The historical developments during the Greek period and again in the modern

European period furnish rich insights into the ways that our present concepts have evolved. A consideration of the historical development also teaches us that problems in science, particularly physics and astronomy, provided the main impetus for the early development of the calculus. We introduce functions that represent physical phenomena, and then use the derivative and integral concepts to provide ways to find areas and volumes of regions and solids, to find velocity and acceleration of moving objects such as cannon balls and planets, and to solve a variety of other practical problems.

SECTION 2 SOME PROBLEMS FROM CALCULUS

Paul Halmos is one of the better-known researchers and expositors of mathematics today. In an article titled "The Heart of Mathematics" in the Aug-Sept 1980 issue of the *American Mathematical Monthly*, he reviewed seven books on problems in mathematics, and stated that "the mathematician's main reason for existence is to solve problems, and that, therefore, what mathematics really consists of is problems and solutions." Halmos claims that the teacher does a disservice to students by only using the lecture approach, without forcing the student to grapple with significant problems and theorems, and to learn how to express solutions verbally and in clear written form.

Edwin E. Moise sounds a similar note in the *Preface to the Teacher* from his book *Introductory Problem Courses in Analysis and Topology*, published by Springer-Verlag in 1982. In this Preface, Moise states his conviction that every undergraduate major should take one or two problem-oriented courses. His approach is to present definitions for the terms usually encountered in analysis, and then to provide a list of statements about these terms. No hint is given as to whether they are true or false. The problems consist of providing proofs for those statements that are true and counterexamples for those that are false.

The typical reader of this book will have had only two or three semesters of intuitive calculus, and even that weak exposure to analysis probably occurred at least a year ago. A good way to begin this more theoretical study of analysis is to clear the "rust" from the mind by considering the following examples. While these examples should all sound familiar to you, they will no doubt be remembered only with a good deal of effort. It is suggested that students work together on these problems during the first week of the course, and perhaps even discuss their ideas during the class period. Working on familiar material in this way should help in the transition to the new and more theoretical ideas that are coming. It is emphasized, however, that some of these examples may be hard to understand, and you should not be discouraged if there is incomplete understanding at this time.

In summary, the purpose of this section is twofold—first, to review some of the main problem types from calculus, and second, to preview what is to come in this book. Many theorems of analysis are illustrated by these examples, and some of the solutions can be simplified or extended by results that are covered later. These examples are intended to demonstrate the range of topics in calculus and to provide motivation to learn the theory underlying the techniques used in these solutions.

Example 1   Use mathematical induction to prove that $2^n \leq n!$ for all $n \geq 4$.

Solution.   In the late 1800s, the Italian mathematician Giuseppe Peano presented a set of five axioms, which he showed could provide a basis for deriving the set $\mathbf{N}$ of natural numbers. His fifth axiom is called the Principle of Mathematical Induction. It asserts that if $P(n)$ is a statement defined for each natural number n, if $P(k)$ is known to be true for a fixed value k, and if $P(n+1)$ is true whenever $P(n)$ is true, then $P(n)$ is true for all $n \geq k$.

In this example, we let $P(n)$ be the statement that $2^n \leq n!$. Then $P(1)$ is the statement that $2^1 \leq 1!$, which is not true. $P(2)$ asserts that $2^2 \leq 2!$, which is also not true. The same fate holds for $P(3)$, but $P(4)$ is true because $2^4 \leq 4!$ is a true statement.

We next need to prove that $P(n+1): 2^{n+1} \leq (n+1)!$ is true whenever $P(n): 2^n \leq n!$ is true. But

$$2^{n+1} = 2 \cdot 2^n \leq 2 \cdot n! \leq (n+1)\, n! = (n+1)!$$

so the desired result is established by mathematical induction.   □

Example 2   Find $\lim_{x \to 0} x \sin \frac{1}{x}$ and $\lim_{x \to \infty} x \sin \frac{1}{x}$.

Solution. $\lim_{x \to 0} x \sin \frac{1}{x} = 0$ because for $x \neq 0$, $|\sin \frac{1}{x}| \leq 1$, and so

$$
\begin{aligned}
0 &\leq \left| x \sin \frac{1}{x} - 0 \right| \\
&\leq \left| x \sin \frac{1}{x} \right| \\
&= |x| \left| \sin \frac{1}{x} \right| \\
&\leq |x|
\end{aligned}
$$

We then appeal to the Squeeze Theorem for limits (see Theorem 10.4), which asserts that as x approaches 0, the quantity $|x \sin \frac{1}{x}|$ must approach

0 because it is "squeezed" between 0 and $|x|$, both of which also approach 0. We observe that the Limit Theorem (Theorem 10.2) cannot be used here since $\lim_{x \to 0} \sin \frac{1}{x}$ does not exist because of oscillation.

The second limit $\lim_{x \to \infty} x \sin \frac{1}{x}$ is indeterminate of the form $\infty \cdot 0$ and can be found using L'Hospital's Rule, since

$$\lim_{x \to \infty} \frac{\sin \frac{1}{x}}{\frac{1}{x}} = \lim_{x \to \infty} \frac{\cos \frac{1}{x} \left(\frac{-1}{x^2} \right)}{\frac{-1}{x^2}} = \cos 0 = 1$$

An alternate method is to use the important limit proved early in calculus that $\lim_{t \to 0} \frac{\sin t}{t} = 1$ and the change of variable $t = \frac{1}{x}$ to show that

$$\lim_{x \to \infty} x \sin \frac{1}{x} = \lim_{t \to 0+} \frac{\sin t}{t} = 1$$

By finding these two limits, we obtain information about the graph of the function $f(x) = x \sin \frac{1}{x}$, namely that f has a removable discontinuity at $x = 0$ (if we define $f(0) = 0$) and also that $y = 1$ is a horizontal asymptote. We could further verify that f is continuous for all x and has zeroes at $x = \pm \frac{1}{n\pi}$ for positive integers n. $\quad \square$

Example 3   Find $f''(0)$ if $f(x) = \begin{cases} x^4 \sin \frac{1}{x} & \text{for } x \neq 0 \\ 0 & \text{for } x = 0 \end{cases}$

Solution.   The usual approach of a calculus student to this type of problem is to calculate $f''(x)$ and then set $x = 0$. While this approach is often successful, we show that it will not work for this problem. By use of the Product Rule, Chain Rule, and the derivative formulas for $\sin x$ and $\cos x$, we find that for $x \neq 0$,

$$f'(x) = 4x^3 \sin \frac{1}{x} - x^2 \cos \frac{1}{x}$$

$$f''(x) = 12x^2 \sin \frac{1}{x} - 6x \cos \frac{1}{x} - \sin \frac{1}{x}$$

While we are able to get $f'(0) = \lim_{x \to 0} f'(x) = 0$ by the Squeeze Theorem approach used in Example 2, that method does not apply to $f''(0)$ because $\lim_{x \to 0} \sin \frac{1}{x}$ does not exist, and so $\lim_{x \to 0} f''(x)$ does not exist either. This proves that f'' is not continuous at 0, but this does not give information as to whether or not $f''(0)$ exists.

We now apply the derivative definition to find

$$f''(0) = \lim_{x \to 0} \frac{f'(x) - f'(0)}{x - 0} = \lim_{x \to 0} \left(4x^2 \sin \frac{1}{x} - x \cos \frac{1}{x} \right) = 0$$

A study of the function $f(x) = x^n \sin \frac{1}{x}$ shows there are infinitely many different "levels" of functions, the first six of which are shown below. A generalization of Example 2 asserts that if a function belongs to level n, it will belong to every level less than n also. And if a function fails to belong to level n, it will not belong to any level greater than n. $f(x) = x^n \sin \frac{1}{x}$ is a function that belongs to the first n levels, but to no others. For example, we showed above that $f(x) = x^4 \sin \frac{1}{x}$ belonged to level 4 but not to level 5. In general, a function f will have a Maclaurin series representation only if it belongs to every one of these infinite number of levels.

Level 1: f is continuous at $x = 0$

Level 2: f is differentiable at $x = 0$

Level 3: f' is continuous at $x = 0$

Level 4: f' is differentiable at $x = 0$

Level 5: f'' is continuous at $x = 0$

Level 6: f'' is differentiable at $x = 0$

$\vdots$ $\square$

Example 4   Apply the Mean Value Theorem for derivatives to $f(x) = \frac{1}{x}$ on $[a, b]$.

Solution.   In order to satisfy the hypotheses for this theorem, we must choose the interval $[a, b]$ so that f will be continuous on the closed interval $[a, b]$ and differentiable on the open interval (a, b). Since $f(x) = \frac{1}{x}$ is continuous and differentiable for all $x \neq 0$, we must choose $[a, b]$ so that $a < b < 0$ or $0 < a < b$.

According to the Mean Value Theorem, there will be at least one value $c \in (a, b)$ so that $f'(c) = \frac{f(b) - f(a)}{b - a}$. For the function $f(x) = \frac{1}{x}$, this equation becomes

$$\frac{-1}{c^2} = \frac{\frac{1}{b} - \frac{1}{a}}{b - a} = \frac{-1}{ab}$$

which simplifies to $c^2 = ab$, or $c = \pm\sqrt{ab}$, where we choose the positive sign if $0 < a < b$ and the negative sign if $a < b < 0$.

The Mean Value Theorem is what we call an existence theorem, in that it asserts the existence of a real number with certain properties, but does not provide a method for finding such a number. We use this theorem to help prove a number of other theorems that have familiar applications, such as L'Hospital's Rule for evaluating limits of indeterminate forms, Taylor's Theorem for finding polynomial approximations of more complicated functions, and the Fundamental Theorem of Calculus for evaluating definite integrals by the use of antiderivatives.   □

Example 5   Find a way to represent the function $F(x) = \int_0^x [t]\,dt$ without the use of an integral sign, where $[t]$ represents the greatest integer function of t. This means that $[t] = n$, where n is the greatest integer for which $n \leq t$.

Solution.   We use an additive property for the definite integral and the fact that we can replace $[t]$ by -1 when $-1 \leq t < 0$, by 0 when $0 \leq t < 1$, by 1 when $1 \leq t < 2$, and so on, in order to solve this problem by cases.

If $0 \leq x < 1$, we have $F(x) = \int_0^x 0\,dt = 0$.

If $1 \leq x < 2$, we have

$$F(x) = \int_0^1 [t]\,dt + \int_1^x [t]\,dt = \int_0^1 0\,dt + \int_1^x 1\,dt = x - 1$$

If $2 \leq x < 3$, we have

$$F(x) = \int_0^1 0\,dt + \int_1^2 1\,dt + \int_2^x 2\,dt$$

$$= 1 + (2x - 4) = 2x - 3$$

You are invited to work a few more cases in order to establish the pattern shown below.

$$F(x) = \begin{cases} \;\;\vdots \\ -2x - 1 & \text{if } -2 \leq x < -1 \\ -x & \text{if } -1 \leq x < 0 \\ 0 & \text{if } 0 \leq x < 1 \\ x - 1 & \text{if } 1 \leq x < 2 \\ 2x - 3 & \text{if } 2 \leq x < 3 \\ 3x - 6 & \text{if } 3 \leq x < 4 \\ \;\;\vdots \end{cases}$$

By observing the graphs of $f(t)$ and $F(x)$ in Figure 2.1, we see that f has jump discontinuities at every integer value, yet is still integrable. We will prove that continuous functions are always integrable, but the converse is not necessarily true. We also see illustrated the theorem that the function defined by an integral is always continuous, even if the integrand is not continuous. However, the integral function will not necessarily be differentiable at a point, unless the integrand is continuous there. The sharp corners in the graph of F show where the function F is not differentiable. If F itself were integrated, the corners would be removed, demonstrating the statement sometimes made that integration is a smoothing process. □

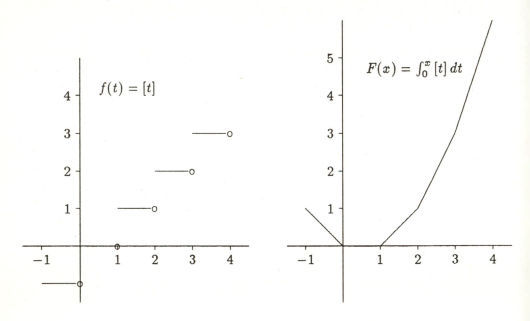

Figure 2.1

Example 6 Evaluate $\int_0^1 x^p \, dx$, where $p \in \mathbf{R}$.

Solution. In order to evaluate this integral, we need to distinguish two possibilities. If $p \geq 0$, then $\int_0^1 x^p \, dx$ is a proper integral because x^p is continuous on $[0, 1]$. Since we know an antiderivative, we use the Fundamental Theorem to write

$$\int_0^1 x^p \, dx \;=\; \left.\frac{x^{p+1}}{p+1}\right|_0^1 \;=\; \frac{1}{p+1}$$

For values of $p < 0$, x^p has an infinite discontinuity at 0, so $\int_0^1 x^p \, dx$ is an improper integral. We evaluate it by considering three subcases.

Case 1: $-1 < p < 0$

$$\int_0^1 x^p \, dx \;=\; \lim_{t \to 0+} \int_t^1 x^p \, dx \;=\; \lim_{t \to 0+} \left. \frac{x^{p+1}}{p+1} \right|_t^1$$

$$=\; \lim_{t \to 0+} \left(\frac{1}{p+1} - \frac{t^{p+1}}{p+1} \right) \;=\; \frac{1}{p+1}$$

and so the integral converges to $\frac{1}{p+1}$.

Case 2: $p = -1$

$$\int_0^1 x^p \, dx \;=\; \lim_{t \to 0+} \left. \ln x \right|_t^1 \;=\; \lim_{t \to 0+} (0 - \ln t) \;=\; +\infty$$

and so the integral diverges to ∞ in this case.

Case 3: $p < -1$

$$\int_0^1 x^p \, dx \;=\; \lim_{t \to 0+} \left(\frac{1}{p+1} - \frac{t^{p+1}}{p+1} \right) \;=\; +\infty$$

since $p + 1 < 0$, and so the integral again diverges to $+\infty$. □

Example 7   Find $F'(x)$ if $F(x) = \int_x^a \sqrt{1 - t^2} \, dt$.

Solution.   One way to do this problem is to use an antiderivative for $\sqrt{1 - t^2}$ in order to find a formula for F without an integral sign, and then to differentiate that expression. A trigonometric substitution would be needed for this, but there is a quicker solution available.

The less familiar half of the Fundamental Theorem of Calculus says that we can find F' by substituting the variable upper limit for t in the integrand if the integrand is continuous at such values. If we interchange the order of the limits of integration, and use the Chain Rule, we obtain the following answer.

$$F'(x) \;=\; \frac{d}{dx} F(x) \;=\; -\frac{d}{dx} \left(\int_a^x \sqrt{1 - t^2} \, dt \right) \;=\; -\sqrt{1 - x^2} \quad \square$$

Example 8   Find the value of $\sum_{n=1}^{\infty} \frac{1}{n^2}$.

Solution. $\sum_{n=1}^{\infty} \frac{1}{n^2}$ is a convergent p-series (as shown in Section 9), so the terms of the sequence of partial sums form an increasing sequence of lower bounds for the value of the series. Thus we have

$$1 \; < \; 1 + \frac{1}{4} \; < \; 1 + \frac{1}{4} + \frac{1}{9} \; < \; \cdots \; < \; \sum_{n=1}^{\infty} \frac{1}{n^2}$$

To get upper bounds, we observe from the graph in Figure 2.2 that improper integrals ($\int_n^{\infty} \frac{1}{x^2} \, dx = \frac{1}{n}$) can be used as follows.

$$\sum_{n=1}^{\infty} \frac{1}{n^2} \; < \; 1 + \int_1^{\infty} \frac{1}{x^2} \, dx \; = \; 2$$

$$< \; 1 + \frac{1}{4} + \int_2^{\infty} \frac{1}{x^2} \, dx \; = \; 1.75$$

$$< \; 1 + \frac{1}{4} + \frac{1}{9} + \int_3^{\infty} \frac{1}{x^2} \, dx \; = \; 1.694\ldots$$

$$\vdots$$

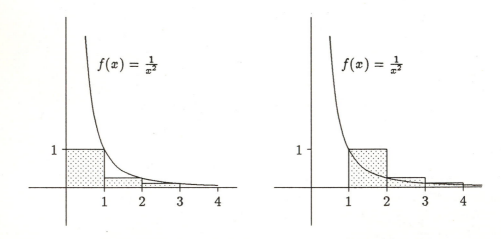

| Figure 2.2 | Figure 2.3 |

A set of lower bounds can also be obtained from the graph by shifting the rectangles one unit to the right, as shown in Figure 2.3, to obtain

$$\int_1^\infty \frac{1}{x^2}\, dx \;<\; 1 + \int_2^\infty \frac{1}{x^2}\, dx \;<\; 1 + \frac{1}{4} + \int_3^\infty \frac{1}{x^2}\, dx \;<\; \cdots < \sum_{n=1}^\infty \frac{1}{n^2}$$

or

$$1 \;<\; 1.5 \;<\; 1.583 \;<\; \cdots < \sum_{n=1}^\infty \frac{1}{n^2}$$

Therefore we know that $1.583 < \sum \frac{1}{n^2} < 1.694$, and by continuing the above process, we can squeeze the value of $\sum \frac{1}{n^2}$ into as small an interval as we wish. By a method such as Fourier series expansions in Chapter 9, we can show that

$$\sum_{n=1}^\infty \frac{1}{n^2} \;=\; \frac{\pi^2}{6} \;\approx\; 1.645 \qquad \square$$

Example 9   Evaluate $\int_0^\infty e^{-x^2}\, dx$.

Solution. $\int_0^\infty e^{-x^2}\, dx$ is also an improper integral, but for a different reason than $\int_0^1 x^p\, dx$ in Example 6. Here the infinite value is in the upper limit instead of in the integrand. Also, this will be a more difficult integral than the one in Example 6 because there is no simple antiderivative for e^{-x^2}. Therefore we will have to resort to an approximation approach, as we did for the infinite series in Example 8, and the reader will see parallels between convergence of infinite series and convergence of improper integrals.

We begin by observing that $\int_0^1 e^{-x^2}\, dx$ is a proper integral because the integrand is continuous on $[0, 1]$. Next, since $e^{-x^2} \le e^{-x}$ for $x \ge 1$, we can compare the two improper integrals as follows.

$$\int_1^\infty e^{-x^2}\, dx \;\le\; \int_1^\infty e^{-x}\, dx \;=\; \left. \frac{-1}{e^x} \right|_1^\infty \;=\; \frac{1}{e}$$

Therefore $\int_0^\infty e^{-x^2}\, dx$ will converge and be equal to the sum

$$\int_0^1 e^{-x^2}\, dx \;+\; \int_1^\infty e^{-x^2}\, dx$$

We have established that $0 < \int_1^\infty e^{-x^2}\, dx < \frac{1}{e}$, and now need to find a value for $\int_0^1 e^{-x^2}\, dx$. We do this below by using a Maclaurin series expansion for the function e^x, and assume that we can replace x by $-x^2$ and integrate term-by-term to obtain

$$e^x = 1 + x + \frac{x^2}{2!} + \cdots + \frac{x^n}{n!} + \cdots$$

$$e^{-x^2} = 1 - x^2 + \frac{x^4}{2!} - \frac{x^6}{3!} + \cdots$$

$$\int_0^1 e^{-x^2}\, dx = \left(x - \frac{x^3}{3} + \frac{x^5}{5\cdot 2!} - \frac{x^7}{7\cdot 3!} + \cdots \right)\Big|_0^1$$

$$= 1 - \frac{1}{3} + \frac{1}{5\cdot 2!} - \frac{1}{7\cdot 3!} + \frac{1}{9\cdot 4!} - \cdots$$

We next use the fact that an alternating series furnishes both upper and lower bounds in order to show that

$$1 - \frac{1}{3} < 1 - \frac{1}{3} + \frac{1}{10} - \frac{1}{42} < \cdots < \int_0^1 e^{-x^2}\, dx < \cdots < 1 - \frac{1}{3} + \frac{1}{10} <$$

$$.667 < .743 < \cdots < \int_0^1 e^{-x^2}\, dx < \cdots < .767 < 1$$

Putting all the above results together, we have

$$0 + .743 < \int_0^\infty e^{-x^2}\, dx < .767 + \frac{1}{e} < 1.135$$

It is possible to obtain better approximations by using more terms of the alternating series, and by using a different function to bound e^{-x^2} on $[1, \infty)$. As one example, xe^{-x^2} has an antiderivative and provides a more accurate upper bound than does e^{-x} on the interval $[1, \infty)$. It would also help to find a lower bound function that has a known antiderivative.

Now that we have done a significant amount of work using results for functions of one real variable to obtain an approximation for $\int_0^\infty e^{-x^2}\, dx$, we next observe that by using double integrals, we can find the value for $\int_0^\infty e^{-x^2}\, dx$. We do this by evaluating the double integral of the function $f(x, y) = e^{-(x^2+y^2)}$ over the first quadrant by changing to an equivalent double integral with polar coordinates. We first represent the value of

$\int_0^\infty e^{-x^2} dx$ by V. Then

$$\int_0^\infty \int_0^\infty e^{-(x^2+y^2)} \, dy \, dx \;=\; \int_0^\infty e^{-x^2} \left(\int_0^\infty e^{-y^2} \, dy \right) dx$$

$$=\; V \int_0^\infty e^{-x^2} \, dx \;=\; V^2$$

Also

$$\int_0^\infty \int_0^\infty e^{-(x^2+y^2)} \, dy \, dx \;=\; \int_0^{\frac{\pi}{2}} \int_0^\infty r\, e^{-r^2} \, dr \, d\theta$$

$$=\; \int_0^{\frac{\pi}{2}} \left. -\frac{1}{2} e^{-r^2} \right|_0^\infty \, d\theta$$

$$=\; \int_0^{\frac{\pi}{2}} \frac{1}{2} \, d\theta \;=\; \frac{\pi}{4}$$

Therefore $V^2 = \frac{\pi}{4}$, so $V = \frac{\sqrt{\pi}}{2}$. In Chapter 6, we will show that this integral is half the value of the Gamma function $\Gamma(x)$ at $x = \frac{1}{2}$. $\square$

Example 10   Show that the sequence $\{s_n\}$ converges, where

$$s_n \;=\; 1 + \frac{1}{2} + \frac{1}{3} + \cdots + \frac{1}{n} - \ln n$$

Solution.   We assume here and throughout the book that $\ln x$ refers to the logarithm function with the base e. There are two key ideas behind the solution to this problem. The first is the Monotone and Bounded Sequence Theorem, which will be proved in Section 6. This theorem states that a sequence which is decreasing and bounded below must converge. It will also be used in Chapter 8 to derive the convergence tests for infinite series. The second key idea is the geometric interpretation of $\ln n$ as the area under the curve $f(x) = \frac{1}{x}$ between $x = 1$ and $x = n$.

We now show that $\{s_n\}$ is a decreasing sequence. This result follows from the definition of s_n and by comparing the two areas shown in Figure 2.4.

$$s_n - s_{n+1} \;=\; \left(1 + \frac{1}{2} + \cdots + \frac{1}{n} - \ln n \right)$$

$$- \left(1 + \frac{1}{2} + \cdots + \frac{1}{n} + \frac{1}{n+1} - \ln(n+1) \right)$$

$$=\; \ln(n+1) - \ln n - \frac{1}{n+1} \;\geq\; 0$$

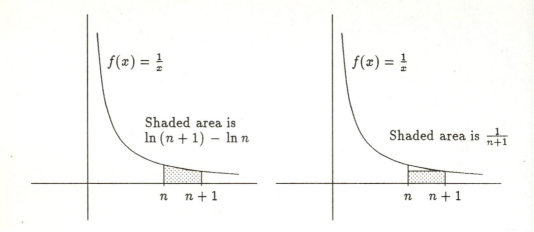

Figure 2.4

We next use another graph of $f(x) = \frac{1}{x}$ in Figure 2.5 on page 17 in order to show that by comparing the area under the rectangles to the area under the curve, we have

$$\left(1 + \frac{1}{2} + \cdots + \frac{1}{n-1}\right) > \int_1^n \frac{1}{x}\, dx = \ln n$$

The shaded portion shows the amount by which the area under the rectangles, which is given by $1 + \frac{1}{2} + \cdots + \frac{1}{n-1}$, exceeds the area under the curve, which is given by $\int_1^n \frac{1}{x}\, dx$. The inequality above implies that the sequence $\{s_n\}$ is bounded below by 0 because

$$s_n = 1 + \frac{1}{2} + \cdots + \frac{1}{n} - \ln n$$

$$= \left(1 + \frac{1}{2} + \cdots + \frac{1}{n-1}\right) - \ln n + \frac{1}{n}$$

$$> 0 + \frac{1}{n} > 0$$

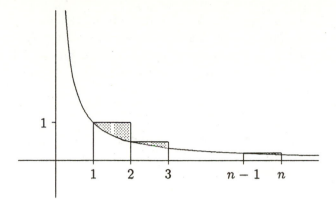

Figure 2.5

At this point, we claim that the sequence must converge and to a value s so that $0 \leq s < s_n$. This is again an existence theorem, which does not provide a method to evaluate the limit. However, it is a well-known value in the mathematical literature, known as Euler's constant, and denoted by the Greek letter γ. It is approximately equal to 0.577, but it is still not known whether γ is a rational or irrational number.   □

Example 11   Find the Maclaurin series expansion for the function

$$f(x) = \frac{x}{e^x - 1}$$

Solution.   The usual method that the student learns in calculus for this type of problem is to calculate several derivatives of f, and look for a pattern, so that an expression may be found for the nth derivative. We can then write

$$f(x) = \sum_{n=0}^{\infty} \frac{f^{(n)}(0)}{n!} x^n$$

This method works very well for functions such as $\sin x$, $\cos x$, e^x, $\sqrt{x + 1}$, and $\ln(x + 1)$.

However, it does not work for functions such as $\tan x$ or $\arctan x$, where the derivatives become increasingly difficult to calculate, and no general pattern emerges. We then seek alternative methods. There is a lovely method for $\arctan x$, since we can begin with the geometric series for $\frac{1}{x+1}$, replace x by x^2 and then integrate term-by-term (see Exercise 10). The best method for $\tan x$ seems to be long division of the series for $\sin x$ by the series for $\cos x$. This is long and tedious, and we do not obtain a simple pattern to express the general term. The function in this problem fits into the same category. The best method seems to be long division. We begin by writing

$$\frac{x}{e^x - 1} = \frac{x}{\left(1 + x + \frac{x^2}{2!} + \frac{x^3}{3!} + \cdots\right) - 1} = \frac{1}{1 + \frac{x}{2!} + \frac{x^2}{3!} + \cdots}$$

and then by the operation of long division,

$$1 + \frac{x}{2} + \frac{x^2}{6} + \frac{x^3}{24} + \cdots \overline{\smash{\big)}\ 1} \quad\quad \overset{\displaystyle 1 - \frac{x}{2} + \frac{x^2}{12} - \frac{x^4}{720} + \cdots}{}$$

This is not the end of the story however, as the Bernoulli numbers are defined to be the numbers B_n so that

$$\frac{x}{e^x - 1} = \sum_{n=0}^{\infty} \frac{B_n}{n!} x^n$$

By comparing these two series, we find that

$$B_0 = 1, \quad B_1 = -\frac{1}{2}, \quad \frac{B_2}{2!} = \frac{1}{12}, \quad \frac{B_3}{3!} = 0, \quad \frac{B_4}{4!} = -\frac{1}{720}, \quad \cdots$$

so that

$$B_0 = 1, \quad B_1 = -\frac{1}{2}, \quad B_2 = \frac{1}{6}, \quad B_3 = 0, \quad B_4 = -\frac{1}{30}, \quad \cdots$$

With more work, it can be shown that $B_{2n+1} = 0$ for all positive integers n, and that $B_6 = \frac{1}{42}$ and $B_8 = -\frac{1}{30}$. These Bernoulli numbers occur in many unexpected places in analysis, such as in the Maclaurin series for $\tan x$ and in the formula for the value of the convergent p-series $\sum_{n=1}^{\infty} \frac{1}{n^{2p}}$. $\square$

By studying these 11 examples, you have reviewed many of the main concepts from the calculus. No doubt, there are places where you still have questions, but most of these should be answered as you work through the rest of the book. In particular, the sections where additional information can be found about the material presented in each example are given below.

Example 1 - Sections 6 and 7
Example 2 - Sections 10, 15 and 22
Example 3 - Sections 15 and 19
Example 4 - Section 20
Example 5 - Sections 25 and 27
Example 6 - Section 28
Example 7 - Section 27
Example 8 - Sections 31, 33 and 36
Example 9 - Sections 29, 32 and 35
Example 10 - Sections 6 and 24
Example 11 - Sections 21 and 35

EXERCISES 2

1. Use mathematical induction to prove that

$$\sum_{i=1}^{n} i^2 = \frac{n(n+1)(2n+1)}{6}$$

2. Verify the formula for $f''(x)$ for $x \neq 0$, as given in Example 3 for $f(x) = x^4 \sin \frac{1}{x}$.

3. Referring to the discussion in Example 3, show that the function F belongs to Level 3 for $x = 0$, but not to Level 4, where

$$F(x) = \begin{cases} x^3 \cos \frac{1}{x} & \text{if } x \neq 0 \\ 0 & \text{if } x = 0 \end{cases}$$

4. Illustrate the Extreme Value Theorem for $f(x) = x^3 - 12x - 4$ on $[-3, 5]$. Illustrate the Intermediate Value Theorem for $f(x) = x^3 - 12x - 4$ on $[-3, 5]$, using the y values of -15 and -25.

5. Apply the Mean Value Theorem for integrals to the function $f(x) = \sin x$ defined on $[0, \pi]$. This result is sometimes called the Theorem of the Mean.

6. In Example 5, verify that $F(x) = 3x - 6$ if $3 \leq x < 4$ and $F(x) = -2x - 1$ if $-2 \leq x < -1$.

7. Express the function $F(x) = \int_{-2}^{x} |t| \, dt$ without an integral sign, following the method in Example 5. Then graph both $f(t) = |t|$ and $F(x)$ to illustrate that the corner in the graph of f is "smoothed out" by the integration process.

8. Work Example 7 by the method of first finding an antiderivative for $\sqrt{1-t^2}$, then using the Fundamental Theorem, and finally calculating F'. Check to see that your answer is the same as the one found by the shorter method in Example 7.

9. Use $\int_{1}^{\infty} x e^{-x^2} \, dx$ to obtain an upper bound for $\int_{1}^{\infty} e^{-x^2} \, dx$. Then find the next two terms in the alternating series given in Example 9 for $\int_{0}^{1} e^{-x^2} \, dx$, and use all these refinements to obtain

$$.7467 \; < \; \int_{0}^{\infty} e^{-x^2} \, dx \; < \; .93143$$

10. Obtain the Maclaurin series for $\arctan x$ using the method described in Example 11. Begin with the geometric series $\frac{1}{1+x} = 1 - x + x^2 - x^3 + \cdots$, replace x by x^2 and then integrate both sides. Next set $x = 1$ to find a well-known approximation for π.

11. In Example 11, carry out the long division to show that

$$\frac{x}{e^x - 1} = 1 - \frac{x}{2} + \frac{x^2}{12} - \frac{x^4}{720} + \cdots$$

Then verify the values given for B_0, B_1, B_2, B_3 and B_4.

12. Use long division as suggested in Example 11 to find the first 4 non-zero terms in the Maclaurin series for $\tan x$. Then verify that these satisfy the formula

$$\tan x \; = \; \sum_{n=1}^{\infty} \frac{(-1)^{n-1} 2^{2n} (2^{2n} - 1) B_{2n}}{(2n)!} x^{2n-1}$$

where B_n represents the nth Bernoulli number as defined in Example 11.

SECTION 3 A BRIEF HISTORY OF CALCULUS

This first chapter is a chapter of introductions. In Section 1, we introduced many theorems and terms that would be covered in the rest of the book. In Section 2, we introduced a variety of problems that are encountered in calculus. We now provide a brief historical introduction to the people and ideas that have shaped the development of calculus. This will furnish a framework for the rest of the book and the more detailed historical comments that occur throughout. In general, the term "calculus" is used to refer to the collection of results about continuity, derivatives, integrals, and series of real-valued functions, while the term "analysis" refers to the careful, theoretical establishment and generalization of these results.

It is helpful to think of the historical development of calculus as occurring in three time periods. The first period begins in the Greek culture during the third century B.C. with Archimedes of Syracuse, and is a time of preparation for the discovery of the calculus by Newton and Leibniz toward the end of the 1600s. The second period, characterized by an extensive development of results of the calculus, extends from the discovery of the calculus by Newton and Leibniz in the years between 1665 to 1676 until the early 1800s. The final period, one of establishing a rigorous foundation to support the results of calculus, begins with Cauchy in 1821 and continues until almost the end of the nineteenth century. At the risk of oversimplification, we might say that calculus was discovered and developed during the second period and analysis was developed during the third period. The time chart in this section shows some of the major individuals who contributed to the development of calculus during these three periods, and should be a helpful reference for you throughout the entire book.

Since the typical calculus student rarely studies the mathematics of the Greek period, you may be surprised to learn how many of our modern concepts can be traced back to the time during the third and fourth centuries B.C. The present-day definition of a real number stems from the theory of ratio and proportion of Eudoxus, while results on conic sections used throughout the calculus were developed by Apollonius. Archimedes was centuries ahead of his time with ingenious summing techniques to find areas of regions and volumes of solids of revolution. In addition, the Greek insistence on rigor and proof, largely neglected until the nineteenth century, has again come into vogue to characterize the nature of modern mathematics, including the analysis studied throughout this book.

After the death of Archimedes in 212 B.C., there was little progress in calculus, until the invention of the printing press at the end of the fifteenth century led to a distribution of Greek mathematical writings in Europe. In addition, the spirit of inquiry of the Renaissance encouraged the earnest study of mathematical questions after an intellectual hibernation of over 1500 years. This era of scientific inquiry accompanied the revival of interest in questions of planetary motion and navigation of the seas to seek new lands and ideas. Both of these required extensive mathematical calculations, and so better calculation techniques were sorely needed. It was still long before the era of the hand-held calculator and the computer, but the discovery of logarithms by John Napier provided the needed tool for those doing lengthy calculations.

The astronomical interests of Galileo and the planetary observations of Tycho Brahe encouraged the formulation of the famous three laws of planetary motion set forth by Johann Kepler. Questions of motion continued to interest people like Blaise Pascal and Pierre de Fermat, who made significant progress toward the calculation of tangent lines to curves. Much of their work would scarcely be recognized by modern students because the symbolism is more complicated than what we now use. There was also a heavy reliance on techniques of Euclidean geometry, and an absence of the concise formulas we use so freely today. One reason for this was that the analytic geometry of Rene Descartes was not published until 1637, and it took time for mathematicians to understand his complicated writings well enough to use the ideas comfortably in their work. Other men such as Bonaventura Cavalieri, John Wallis, Isaac Barrow, and James Gregory, went beyond the techniques of Archimedes to find areas for more general curves than the parabola and circle.

When Isaac Newton and Gottfried Leibniz began their study of tangent lines and areas in the period from 1665 to 1676, they had available to them a host of successfully solved specific problems, but no general methods or techniques. It was their genius to independently discover that the two problems were closely related. They found general methods for calculating derivatives (Newton called these fluxions, while Leibniz used the term differential), which would give slopes of tangent lines, or velocities and accelerations of moving objects. Then they showed the problem of areas (or integration) could be solved by a process of inverse differentiation.

During the next 150 years, the contemporaries and successors of Newton and Leibniz built upon their discovery a massive structure of results with applications to a wide range of physical problems. Brook Taylor and Colin Maclaurin were the best known of the English mathematicians, who followed the direction laid out by their countryman Newton. The continental mathematicians adopted the more useful symbolism of Leibniz. The main mathematicians here include the brothers James and John Bernoulli, followed in time by Leonhard Euler, and then Joseph Lagrange.

Sporadically throughout the eighteenth century, concern was expressed about the concept of infinitesimals on which the calculus results were based. This philosophical idea provided one way to handle the "infinitely small" entities needed for the definitions of the basic concepts. The most well-known criticism of these came from Bishop Berkeley in 1734 through his writing of *The Analyst*. However, none of the proposed answers to these problems proved effective, and most mathematicians ignored the philosophical problems anyway.

Towards the end of the eighteenth century and into the beginning of the nineteenth century, new concerns arose over the foundations of calculus. All the results of calculus had been established very intuitively, so there was no firm basis under these theorems. The work of Joseph Fourier with his use of trigonometric series for the study of heat transfer led to new kinds of functions that strained the fragile theory of continuity, derivatives, and integrals to the breaking point.

From 1820 until about 1890, a number of gifted mathematicians responded to this need by gradually establishing a firm basis for calculus in the spirit of the axiomatic method that had been characteristic of the Greek approach so long ago. Niels Abel from Norway, Bernhard Bolzano from Prague, Augustin-Louis Cauchy from France, and Lejeune Dirichlet from Germany were some of the earliest to undertake this task, with the majority of the credit belonging to Cauchy. In his university lectures in 1821, he set forth the idea of a limit, with its concept of small tolerances that we now denote by ϵ and δ, as the basis on which the rest of the calculus could be built. He showed how to define continuity, the derivative, the definite integral, and convergence of sequences and series in terms of this limit concept, and obtained a number of standard theorems complete with proofs, most of which were correctly done, even by modern standards. We essentially retrace his steps in this book, hopefully avoiding his errors, and adding comments and results where appropriate.

As Cauchy's concept of the integral had some limitations, George Riemann put this definition into its present form in the 1850s, and also provided an important necessary and sufficient condition for integrability. His definition was in a form that permitted Henri Lebesgue to successfully generalize it around 1900, when the theory of sets from Georg Cantor was available to provide the needed framework for such an extension. This infinite set theory of Cantor led to the fields of measure theory and point-set topology that have been fruitful in twentieth century work. Cauchy and Riemann also extended the calculus concepts to functions of a complex variable, and developed this area into a complete and useful theory.

One final development needs to be mentioned. As the results of the calculus, being built on the foundation of limits, were organized into the theoretical structure of analysis, the realization grew that the real numbers themselves, on which limits depended, had no true foundation. In the period from 1860 to 1880, the German mathematicians Karl Weierstrass, Richard Dedekind, and Georg Cantor provided different but equivalent ways to obtain the complete ordered field of real numbers. It seemed that analysis was finally beyond reproach or attack, but we should always be careful when such complacency occurs. Such developments as the discovery of paradoxes in set theory touched off, in the opening decades of the twentieth century, a search to find the proper philosophy of mathematics that could serve as a correct foundation for the real numbers. In more recent years, the long-neglected entity of infinitesimals has been resurrected and used as an alternate basis for the calculus. So the final chapter on the history of calculus or analysis has not yet been written.

EXERCISES 3

1. What are the characteristics of the three periods for the development of calculus and analysis, as discussed in this section?

2. From the chart of the 27 mathematicians on page 25, make a list of all the names that are familiar to you, including the facts you know about them. For example, who would you associate with the Cartesian coordinate system, the triangular array of coefficients for the binomial expansion, or the laws of planetary motion?

3. What did Greek culture contribute to the development of calculus?

4. What was the special contribution of Newton and Leibniz?

5. After Newton and Leibniz, name three other mathematicians that you feel were important in the development of calculus, and tell why.

6. How would you distinguish between calculus and analysis?

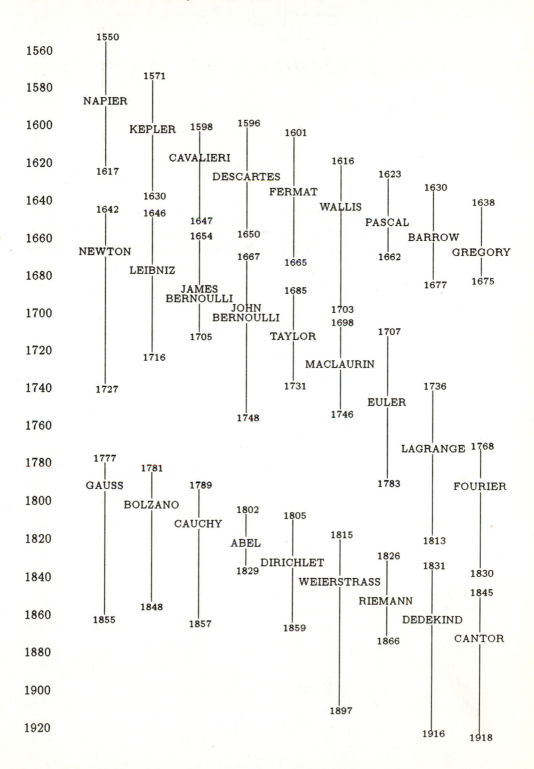

2

Sequences and Basic Axioms

SECTION 4   CONVERGENT SEQUENCES

We begin our study of analysis by using sequences to introduce the limit concept, which is the foundational idea for the entire book. In most calculus books, the sequence idea is not presented until the middle of the book when infinite series are introduced. The limit concept and the ϵ-δ symbolism are presented at the beginning of such books in the context of functions rather than sequences.

One reason for beginning with sequences in this book is that the rigorous epsilon proofs that we give are usually understood more readily in the context of sequences than in that of functions of a real variable. We will treat the epsilon theory for functions in Chapter 3. Another reason for beginning with sequences is that they provide a good vehicle for developing the Bolzano-Weierstrass Theorem, which is an important theoretical result for the proofs of a number of theorems encountered throughout the course. So it is not that we must begin the book with sequences, but rather that they provide the best transition into the subject matter of analysis.

Definition 1   A *sequence of real numbers* is a real-valued function whose domain is the set $\mathbf{N}$ of natural numbers.

We use a symbol such as x_n to represent the range value $f(n)$ and the symbol $\{x_n\}_{n=1}^{\infty}$ will represent the sequence itself. Some books denote a sequence by the symbol (x_n). A sequence can be considered as an ordered array of real numbers, or as a special kind of function. Two examples of sequences are

$$\{x_n\} = \{(-1)^n\} = \{-1, +1, -1, +1, \ldots\}$$

and

$$\{x_n\} = \left\{\frac{1}{n^2}\right\} = \left\{1, \frac{1}{4}, \frac{1}{9}, \frac{1}{16}, \ldots\right\}$$

Sequences can have other sets of indices, such as $\{x_n\}_{n=0}^{\infty}$ or $\{x_n\}_{n=k}^{\infty}$. However, since most sequences are indexed from 1 to ∞, we will use $\{x_n\}$ to mean $\{x_n\}_{n=1}^{\infty}$. In this chapter, the values for the x_n terms will be real numbers, but we can also have sequences of complex numbers, sequences of functions, sequences of sets, or sequences of derivatives or integrals. We will consider some of these possibilities later.

Throughout analysis, the Greek letter epsilon (ϵ) is used to represent a small positive number. We use it here to formally express the idea that the sequence terms x_n get close to the limit x if the expression $|x_n - x|$ gets close to 0, or if for large enough values of n, the expression $|x_n - x|$ can be made less than any positive tolerance. We use the symbol ϵ to represent this positive tolerance.

The most important kinds of sequences of real numbers that we consider are those that converge in the following sense.

Definition 2   The sequence $\{x_n\}$ is said to *converge to* x if, for every $\epsilon > 0$, there is a positive integer N so that $|x_n - x| < \epsilon$ holds for all $n > N$. In this case, we write $\lim_{n \to \infty} x_n = x$, and often shorten this to $\lim x_n = x$. We sometimes use the symbol N_ϵ or $N(\epsilon)$ to indicate that the value of N depends on the value of ϵ. The value x is called the *limit* of the sequence.

Comment.   Modern definitions, such as sequence convergence in Definition 2, are usually expressed formally by the use of symbols, such as $|x_n - x| < \epsilon$. The early forms of these definitions were almost always expressed using words only. For example, Cauchy wrote that a sequence converges

> when the successive values attributed to a variable approach indefinitely a fixed value so as to end by differing from it by as little as one wishes, this last fixed value is called the *limit* of all the others.[1]

You may recall several places in your calculus course where the concept of a convergent sequence was used, at least implicitly, to define an important idea from calculus. Four of these are now listed.

---

[1] "Infinitely Small Quantities in Cauchy's Textbook" by Detlef Langwitz, *Historia Mathematica*, August 1987, p. 263.

1. A sequence that may be used to define the definite integral $\int_a^b f(x)\,dx$, where f is nonnegative, is obtained by dividing the interval $[a, b]$ into n equal parts, and approximating the area under the graph of $y = f(x)$ by the sum of the areas of n rectangles, where $\Delta x = \frac{b-a}{n}$ is the length of the base of each rectangle, and $x_i = a + i\Delta x$ is the right endpoint of the ith subinterval. Then $S_n = \sum_{i=1}^n f(a + i\Delta x)\Delta x$, and $\int_a^b f(x)\,dx$ is defined as $\lim_{n\to\infty} S_n$.

2. In order to present the concept of the improper integral $\int_a^\infty f(x)\,dx$, we begin with the sequence $\{s_n\}$ where $s_n = \int_a^n f(x)\,dx$. The number s_n is well-defined if f is continuous on $[a, n]$. We then define $\int_a^\infty f(x)\,dx$ as the limit of the sequence $\{s_n\}$.

3. To obtain one of the most useful sequences, we define the value of an infinite series as the limit of the sequence of partial sums. For example, $\sum_{i=1}^\infty \frac{1}{i^2}$ is defined as the limit of the sequence $\{s_n\}$ where

$$s_n = \sum_{i=1}^n \frac{1}{i^2} = 1 + \frac{1}{4} + \cdots + \frac{1}{n^2}$$

Similarly, the geometric sequence $\{s_n\}$ where $s_n = 1 + r + r^2 + \cdots + r^n$ is shown in Section 9 to converge to $\frac{1}{1-r}$ if $|r| < 1$. In particular, if $r = \frac{1}{2}$, then

$$\lim_{n\to\infty} s_n = \lim_{n\to\infty} \sum_{k=0}^n \left(\frac{1}{2}\right)^k = \sum_{k=0}^\infty \left(\frac{1}{2}\right)^k = \frac{1}{1 - \frac{1}{2}} = 2$$

4. The specific sequence $\{s_n\} = \{(1 + \frac{1}{n})^n\}$ is used in connection with logarithmic and exponential functions. It converges to the well-known value, which is represented by the letter e, and is approximately equal to 2.71828.

   The intuitive idea behind convergent sequences is that the terms of a sequence get arbitrarily close to some real number x, as the value of n gets increasingly large. In a pictorial sense, if we draw an ϵ-band about the value x, then for all values of n beyond some positive integer N, the sequence terms x_n will lie in this band, which is also referred to as an ϵ-interval about x, or an ϵ-neighborhood of x (see Figure 4.1). The condition that $|x_n - x| < \epsilon$ for all $n > N$ is often informally expressed by the statement that the sequence $\{x_n\}$ is *eventually* in each ϵ-interval about x.

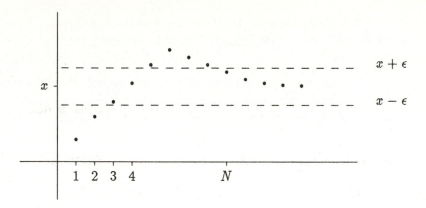

Figure 4.1

Under this definition, the sequence $\{\frac{1}{n^2}\}$ converges to zero, while the sequence $\{(-1)^n\}$ does not converge. Sequences that do not converge are said to *diverge*. The sequence $\{(-1)^n\}$ diverges because the terms oscillate between the values of -1 and $+1$, while the sequence $\{\ln n\}$ is said to diverge to $+\infty$, because the terms of this sequence exceed all positive bounds. Do you think the sequence $\{\frac{2n}{n+2}\}$ converges or diverges?

We now illustrate the definition of convergence by applying it to several examples of sequences.

Example 1   Use Definition 2 to show that the sequence $\{x_n\}$ converges to 2, where $x_n = \frac{2n}{n+2}$.

Solution.   We observe that

$$|x_n - x| = \left| \frac{2n}{n+2} - 2 \right| = \left| \frac{-4}{n+2} \right| = \frac{4}{n+2} < \epsilon$$

will hold if $n + 2 > \frac{4}{\epsilon}$, or if $n > \frac{4}{\epsilon} - 2$. We can let $N_\epsilon = [\frac{4}{\epsilon} - 2]$ to satisfy the conditions of the definition. We use the greatest integer symbol because we want N_ϵ to be a positive integer. Observe that if $\epsilon > \frac{4}{3}$, then $[\frac{4}{\epsilon} - 2]$ will not be a positive integer. Remember the convention that ϵ is an arbitrarily small tolerance, such as .00001. Sometimes the above problem can be avoided by defining $N_\epsilon = \max\{1, [\frac{4}{\epsilon} - 2]\}$.

Now if $n > N_\epsilon$, then $n > \frac{4}{\epsilon} - 2$, which is equivalent to $|x_n - x| < \epsilon$. In particular, $N_{.01} = 398$, which means that all the terms after x_{398} will be between 1.99 and 2.01. Observe that $x_{398} = 1.99$, so that $|x_{398} - 2| = .01$ and $x_{399} \approx 1.9900249$, so that $|x_{399} - 2| < .01$. $\square$

Following the practice begun in Section 2, the symbol $\square$ will be used to indicate the end of an example. The symbol $\blacksquare$ will be used to indicate the completion of the proof of a theorem, lemma, or corollary.

Example 2   Use Definition 2 to show that $\lim_{n \to \infty} \frac{3n+1}{2n^2+3} = 0$.

Solution.   Because of the way n occurs in the expression

$$| x_n - x | = \left| \frac{3n+1}{2n^2+3} - 0 \right|$$

we are unable to solve for n in terms of ϵ as we did in Example 1. By use of the inequality

$$\frac{3n+1}{2n^2+3} < \frac{3n+n}{2n^2} = \frac{2}{n} < \epsilon$$

we will have $|x_n - x| < \epsilon$ if $n > N_\epsilon = [\frac{2}{\epsilon}]$. For $\epsilon = .01$, $N_{.01} = 200$ and we see that $x_{201} \approx .00747$ does differ from 0 by an amount less than .01. However, there are also many earlier terms with this property.

An alternate approach uses the inequality

$$\frac{3n+1}{2n^2+3} < \frac{3n+3}{2n^2-2} = \frac{3(n+1)}{2(n+1)(n-1)} = \frac{3}{2(n-1)}$$

and now $|x_n - x|$ will be less than ϵ if we define $N_\epsilon = [\frac{3}{2\epsilon} + 1]$. In this case $N_{.01} = 151$, which is a more "precise" choice than $N = [\frac{2}{\epsilon}]$. Since $x_{150} \approx .01002$, we apparently now have the smallest possible value for N. We emphasize that Definition 2 only requires that we find a value for N that works—it does not need to be the smallest such value. $\square$

Example 3   Apply Definition 2 to $\lim_{n \to \infty} \frac{1-n^2}{2n^2+3}$.

Solution.   We first find the value of this convergent sequence by using the results in Theorem 4.1 to show that

$$\lim_{n \to \infty} \frac{1-n^2}{2n^2+3} = \lim_{n \to \infty} \frac{\frac{1}{n^2} - 1}{2 + \frac{3}{n^2}} = -\frac{1}{2}$$

We then write

$$\left| \frac{1-n^2}{2n^2+3} - \left(-\frac{1}{2}\right) \right| = \left| \frac{2 - 2n^2 + 2n^2 + 3}{2(2n^2+3)} \right| = \frac{5}{2(2n^2+3)}$$

which will be less than ϵ if we choose $n > \sqrt{\frac{5-6\epsilon}{4\epsilon}}$. Therefore we define $N_\epsilon = \left[\sqrt{\frac{5-6\epsilon}{4\epsilon}} \right]$. Clearly $5 - 6\epsilon$ must be non-negative, so $\epsilon \leq \frac{5}{6}$ must hold in order for N_ϵ to be well-defined.

A simpler, but less precise, expression could be found for N if we insert other terms between $\frac{5}{2(2n^2+3)}$ and ϵ, such as

$$\frac{5}{2(2n^2+3)} < \frac{5}{2(2n^2)} < \frac{16}{4n^2} < \epsilon$$

In this case, $N_\epsilon = [\frac{2}{\sqrt{\epsilon}}]$, and no restriction such as $\epsilon \leq \frac{5}{6}$ is needed. □

In case you feel that this ϵ–N_ϵ procedure is somewhat suspicious, we next illustrate that it does not work when it is applied to an incorrect limit value.

Example 4   If we incorrectly assume in Example 1 that $\lim \frac{2n}{n+2} = 3$, instead of the correct value of 2, then for all n,

$$\left| \frac{2n}{n+2} - 3 \right| = \left| \frac{-n-6}{n+2} \right| = \frac{n+6}{n+2} > 1$$

and so $|x_n - x|$ can never be made less than any value of $\epsilon < 1$. □

Example 5   If $\lim_{n\to\infty} x_n = x$ where $x \geq 0$, and $x_n \geq 0$ for all n, then $\lim_{n\to\infty} \sqrt{x_n} = \sqrt{x}$.

Solution.   Let $\epsilon > 0$ be given. If $x > 0$, $\sqrt{x}\,\epsilon$ is a positive number, so by Definition 2, there is some N so that $|x_n - x| < \epsilon\sqrt{x}$ for all $n > N$. Then

$$\left| \sqrt{x_n} - \sqrt{x} \right| = \left| \frac{x_n - x}{\sqrt{x_n} + \sqrt{x}} \right| \leq \frac{1}{\sqrt{x}} |x_n - x| < \frac{1}{\sqrt{x}} \epsilon\sqrt{x} = \epsilon$$

In the case where $x = 0$, there is some N so that $|x_n - 0| < \epsilon^2$ for all $n > N$. Then $|\sqrt{x_n} - 0| < \epsilon$. □

At this point, we come to the first main theorem of the course—the Limit Theorem for sequences. Stated in language commonly used in abstract algebra, we ask whether the set of all convergent sequences of real numbers is closed under the operations of addition, multiplication, and division. We will discuss the same question in Chapter 3 with limits of functions and have a much expanded result. We will also ask this question of closure for sums, products, and quotients of continuous functions in Chapter 4, differentiable functions in Chapter 5, and integrable functions in Chapter 6. Since this is not only our first theorem, but is clearly a very general and important result, we provide a careful proof for it.

In most epsilon proofs in analysis, we use the triangle inequality, which is the statement that $|a + b| \leq |a| + |b|$ for all real numbers a and b. See Appendix 2 if needed.

THEOREM 4.1 (Limit Theorem for Sequences)

If the sequences $\{x_n\}$ and $\{y_n\}$ converge to x and y respectively, then the following three statements are true.

(1) The sequence $\{x_n + y_n\}$ converges to $x + y$.

(2) The sequence $\{x_n y_n\}$ converges to xy.

(3) The sequence $\{\frac{x_n}{y_n}\}$ converges to $\frac{x}{y}$, if $y \neq 0$ and $y_n \neq 0$ for all n.

Proof of (1). This proof is very straightforward, and depends only on the definition of convergence and the triangle inequality. For some of the first proofs in this book, each step is given on a separate line to help you see the chain of reasoning more clearly.

Let $\epsilon > 0$ be given.

By Definition 2, there is some N_1 so that $|x_n - x| < \frac{\epsilon}{2}$ if $n > N_1$.

Similarly, there is some N_2 so that $|y_n - y| < \frac{\epsilon}{2}$ if $n > N_2$.

Let $N = \max\{N_1, N_2\}$.

For $n > N$, we have

$$|(x_n + y_n) - (x + y)| \;=\; |(x_n - x) + (y_n - y)|$$

$$\leq \; |x_n - x| + |y_n - y|$$

$$< \; \frac{\epsilon}{2} + \frac{\epsilon}{2} = \epsilon$$

Notice that for $n > N$, we have $n > N_1$ and $n > N_2$, so that both inequalities at the start of the proof apply. This completes the proof of (1), since Definition 2 is satisfied for the sequence $\{x_n + y_n\}$.   ■

Comment.   Before we present a formal proof of (2), we will first do some informal groundwork. Most proofs are usually done this way first, even by experienced mathematicians. After the outline of the proof is discovered by this process, a careful formal proof is constructed, and this final work is usually all that the reader sees in print. It is therefore possible to draw the false conclusion that all mathematical proofs originally come in a polished form, such as the proof of (1) above.

To prove (2), we first observe the need to find a value for N so that $|x_n y_n - xy| < \epsilon$ for all $n > N$. Since $\{x_n\}$ converges to x and $\{y_n\}$ converges to y, we are able to make $|x_n - x|$ and $|y_n - y|$ as small as we wish or need.

In order to bring the expressions $|x_n - x|$ and $|y_n - y|$ into play, we use a common device in analysis proofs of adding 0, in the form $0 = -a + a$. For this problem, we could use either $-x_n y + x_n y$ or $-y_n x + y_n x$. Using the first choice, we have

$$|x_n y_n - xy| \;=\; |x_n y_n - x_n y + x_n y - xy|$$

$$=\; |x_n(y_n - y) + y(x_n - x)|$$

$$\leq\; |x_n| \cdot |y_n - y| + |y| \cdot |x_n - x|$$

This last expression will be less than ϵ, if we can make $|x_n| \cdot |y_n - y| < \frac{\epsilon}{2}$ and $|y| \cdot |x_n - x| < \frac{\epsilon}{2}$. Since $|x_n - x|$ can be made arbitrarily small, we first choose to make $|x_n - x| < \frac{\epsilon}{2|y|}$. This will work unless $y = 0$. To avoid this problem, we decide to make $|x_n - x| < \frac{\epsilon}{2|y|+1}$. Why will this now work regardless of the value for y?

We might next be tempted to make $|y_n - y| < \frac{\epsilon}{2|x_n|}$ in a similar way, but this will not work because $|x_n|$ is a variable term and we need a fixed constant. We are therefore led to prove Lemma 4.2 first, which provides the constant bound that we need. Then we organize all the above work into a careful proof of (2). It is hoped that you will gain helpful insight into the method of proof discovery through this discussion, and will learn to use similar techniques.

Definition 3   The sequence $\{x_n\}$ is *bounded above* if there is some real number t, so that $x_n \leq t$ for all n, and *bounded below* if $x_n \geq t$ for all n. The sequence $\{x_n\}$ is *bounded* if it is bounded above and below. This also means that there is some value $B > 0$, so that $|x_n| \leq B$ for all $n \in \mathbf{N}$.

LEMMA 4.2 (Convergent sequences are bounded)

Let the sequence $\{x_n\}$ converge to x. Then $\{x_n\}$ is bounded.

Comment.   For some reason, this simple proof is difficult for many students to understand. We begin by asking if x could be the desired bound. However, this will not be true because all the x_n terms could be greater than x. Can you give an example of this?

Since all the x_n terms must eventually get into every interval about x, we could use the interval $(x - 1, x + 1)$ and let $B = x + 1$. This would be a better choice, but even though $x - 1 < x_n < x + 1$ holds for all $n > N$, the terms before x_N could be outside this interval. In addition, we need to handle the case when $x < 0$. For if $x = -3$, it would not be true that $|x_n| < x + 1 = -2$. We therefore are led to the following proof.

Proof of Lemma 4.2.

There is a positive integer N so that $|x_n - x| < 1$ for all $n > N$. Since $|x_n| - |x| \leq |x_n - x|$ by Exercise 17a, it follows that $|x_n| < |x| + 1$ for $n > N$. Let $B = \max\{|x_1|, |x_2|, \ldots, |x_N|, |x| + 1\}$. Then $|x_n| \leq B$ for all n, and the proof is complete. ∎

We are now able to organize the above discussion into a formal proof of result (2) in Theorem 4.1.

Proof of (2).

Let $\epsilon > 0$ be given. By Lemma 4.2, there is a bound B so that $|x_n| \leq B$ for all n. By Definition 2, there is some N_1 so that $|y_n - y| < \frac{\epsilon}{2B}$ for all $n > N_1$. Similarly, there is some N_2 so that $|x_n - x| < \frac{\epsilon}{2|y|+1}$ for all $n > N_2$. Let $N = \max\{N_1, N_2\}$ and suppose $n > N$. Since n is greater than both N_1 and N_2, the above inequalities on $|x_n - x|$ and $|y_n - y|$ hold, and so

$$
\begin{aligned}
|x_n y_n - xy| &= |x_n y_n - x_n y + x_n y - xy| \\[2ex]
&= |x_n(y_n - y) + y(x_n - x)| \\[2ex]
&\leq |x_n| \cdot |y_n - y| + |y| \cdot |x_n - x| \\[2ex]
&< B \cdot \frac{\epsilon}{2B} + |y| \cdot \frac{\epsilon}{2|y| + 1} \\[2ex]
&< \frac{\epsilon}{2} + \frac{\epsilon}{2} = \epsilon \quad \blacksquare
\end{aligned}
$$

Comment. We have two options for a proof of result (3). The first way is to work intuitively with $\left| \frac{x_n}{y_n} - \frac{x}{y} \right|$ as we did with result (2), until the proper bounds for $|x_n - x|$ and $|y_n - y|$ are discovered. The result stated in Lemma 4.3 below will also be needed. This approach is left as an exercise for you to do.

A second and shorter approach is to first prove that if $\lim y_n = y \neq 0$, then $\lim_{n \to \infty} \frac{1}{y_n} = \frac{1}{y}$, and then to put this result together with result (2) to obtain the proof of (3). This approach is shown below.

LEMMA 4.3   If $\lim_{n \to \infty} y_n = y \neq 0$, then there is a positive integer N, so that $|y_n| \geq \frac{|y|}{2} > 0$ for $n > N$.

Proof. This is left as an exercise. This lemma states that not only are all the y_n terms eventually non-zero, they are actually bounded away from zero.

It might be instructive to combine Lemmas 4.2 and 4.3 into the following form. It is based on the idea that a convergent sequence is eventually in every interval about the limit value. So if the sequence $\{x_n\}$ converges to x, then the sequence $\{|x_n|\}$ converges to $|x|$ (see Exercise 5), and so the terms of $\{|x_n|\}$ are eventually in each of the following intervals: $(|x| - 1, |x| + 1)$, $\left(\frac{|x|}{2}, \frac{3|x|}{2} \right)$, and $\left(\frac{|x|}{2}, |x| + 1 \right)$. However it is expressed, the key idea is that the terms of a convergent sequence eventually are contained in any interval about the limit value x.

LEMMA 4.4   Let the sequence $\{x_n\}$ converge to x. Then there is some N so that $|x|/2 \leq |x_n| \leq |x| + 1$ for all $n > N$.

Proof.   This is left as an exercise.

You will find it instructive to do the informal work first that will make the following formal proof understandable.

LEMMA 4.5   If $\lim_{n \to \infty} y_n = y \neq 0$, and $y_n \neq 0$ for all n, then $\lim_{n \to \infty} \frac{1}{y_n} = \frac{1}{y}$.

Proof.   Let $\epsilon > 0$ be given. By Lemma 4.3, there is some N_1 so that

$$|y_n| \geq \frac{|y|}{2} \quad \text{or} \quad \frac{1}{|y_n|} \leq \frac{2}{|y|} \quad \text{for all } n > N_1$$

By Definition 2, there is some N_2 so that for all $n > N_2$,

$$|y_n - y| = |y - y_n| < \frac{\epsilon |y|^2}{2}$$

Let $N = \max\{N_1, N_2\}$ and choose $n \geq N$. Then

$$\left| \frac{1}{y_n} - \frac{1}{y} \right| = \left| \frac{y - y_n}{y_n y} \right| = \frac{1}{|y_n|} \frac{1}{|y|} |y - y_n| < \frac{2}{|y|} \frac{1}{|y|} \frac{\epsilon |y|^2}{2} = \epsilon$$

and the proof is complete.   ■

We are now able to give a proof for result (3) of Theorem 4.1.

Proof of (3).   By Result (2) of Theorem 4.1 and Lemma 4.5,

$$\lim \frac{x_n}{y_n} = \lim \left(x_n \cdot \frac{1}{y_n} \right) = \left(\lim x_n \right) \cdot \left(\lim \frac{1}{y_n} \right) = x \cdot \frac{1}{y} = \frac{x}{y} \quad ■$$

THEOREM 4.6   If $\{x_n\}$ and $\{y_n\}$ are two convergent sequences with $x_n \leq y_n$ for all $n > k$, for some $k \in \mathbf{N}$, then $\lim x_n \leq \lim y_n$.

Proof.   This is left as an exercise.

EXERCISES 4

1. (a) If $\{x_n\} = \{\sqrt{n+1} - \sqrt{n}\}$, show that $\lim_{n \to \infty} x_n = 0$.

   (b) What can be said about the convergence of the sequence $\{\sqrt{n}\, x_n\}$?

2. Find the first 5 terms of the sequence

$$\{x_n\} = \left\{ \frac{1}{n^2} + \frac{2}{n^2} + \cdots + \frac{n}{n^2} \right\}$$

   Then find $\lim_{n \to \infty} x_n$.

3. Apply Definition 2 to the following sequences. This means you should find an appropriate formula for N.

   (a) $\{x_n\} = \{\frac{1}{n^2+1}\}$

   (b) $\{x_n\} = \{\frac{2n^2-3}{n^3+1}\}$

   (c) $\{x_n\} = \{\frac{4n+1}{2n+5}\}$

4. (a) Apply Definition 2 to $\{x_n\} = \{\frac{3n^2+1}{2n^2+3}\}$.

   (b) Find the values for $N(.1)$ and $N(.01)$.

5. (a) Use Definition 2 to prove that if $\{x_n\}$ converges to x, then $\{|x_n|\}$ converges to $|x|$.

   (b) Give an example to show that the converse result is not true.

6. (a) Find divergent sequences $\{x_n\}$ and $\{y_n\}$ so that $\{x_n + y_n\}$ converges.

   (b) If $\{x_n\}$ converges and $\{y_n\}$ diverges, could $\{x_n + y_n\}$ converge?

7. Find divergent sequences $\{x_n\}$ and $\{y_n\}$ so that $\{x_n y_n\}$ converges.

8. Form an intuitive proof for result (3) of Theorem 4.1, following the one done for result (2) in the text.

9. Prove Lemma 4.3.

10. Prove Lemma 4.4.

11. Give a proof in the form of the one provided for Theorem 4.1 that if $\lim x_n = x$ and $\lim y_n = y$, then $\lim(ax_n + by_n) = ax + by$, for constants a and b.

12. Prove that if $\{x_n\}$ converges to x, then $\{\sqrt[3]{x_n}\}$ converges to $\sqrt[3]{x}$.

13. Prove that if $\{x_n\}$, $\{y_n\}$ and $\{z_n\}$ converge to x, y, and z respectively, then $\{x_n y_n z_n\}$ converges to xyz.

14. Prove that if $\{x_n\}$ converges to x, then $\{x_n{}^k\}$ converges to x^k, for all natural numbers k. Hint: Use mathematical induction.

15. Find $\lim_{n \to \infty} c^n$. Consider cases, depending on the value of c.

16. Prove that if $\{x_n\}$ converges to 0, and $\{y_n\}$ is bounded, then $\{x_n y_n\}$ converges to 0.

17. (a) Use the triangle inequality to prove that $|a| - |b| \leq |a - b|$ for all real numbers a and b. Begin with the formula $a = (a - b) + b$.

    (b) Prove that $\left| |a| - |b| \right| \leq |a - b|$.

18. Prove Theorem 4.6.

19. Prove that if the sequence $\{x_n\}$ converges, and the sequence $\{y_n\}$ diverges to $+\infty$, then the sequence $\{\frac{x_n}{y_n}\}$ converges to 0.

20. Prove that $\lim_{n \to \infty} \frac{x^n}{n!} = 0$ for all $x \in \mathbf{R}$.

21. Use mathematical induction to prove that $\left| \sum_{i=1}^{n} x_i \right| \leq \sum_{i=1}^{n} |x_i|$.

22. Prove that the limit of a convergent sequence is unique. You should assume that $\lim_{n \to \infty} x_n = x$ and $\lim_{n \to \infty} x_n = y$, and then prove that $x = y$.

SECTION 5   SOME RESULTS ABOUT SETS

In the previous section, we presented a definition for the convergence of a sequence and saw how to apply it to specific examples and to the proof of the Limit Theorem. We also saw that while all convergent sequences must be bounded, not all bounded sequences must converge. Our next results deal with conditions under which a bounded sequence will converge. In order to prove these results in Sections 6 and 7, we need some properties of the order relation for the set **R** of real numbers. You are encouraged to refer to Appendices 1 and 2 at this time, as needed, for a brief review of set notation and the structure of the set of real numbers.

Definition 1   The number b is an *upper bound* for the set S if $x \leq b$ holds for all $x \in S$. The number u is a *least upper bound* for S if u is an upper bound for S, and $u \leq b$ for all upper bounds b of S.

Comment on uniqueness.   While a set may have many upper bounds, it cannot have more than one least upper bound. For suppose u and v are least upper bounds for the set S. Since u is a least upper bound for S and v is an upper bound for S, then $u \leq v$. Reversing the roles of u and v, we obtain $v \leq u$, and so $u = v$. As a result, we can speak of *the* least upper bound for a set S, and we can denote it by lub S. In recent years, a more commonly used symbol is sup S, where sup is the abbreviation for *supremum*. If a non-empty set S has no upper bound, we write sup $S = $ lub $S = +\infty$.

Example 1   The number b is the least upper bound for the intervals (a, b) and $[a, b]$. It is not necessary that the least upper bound for S be an element of S.   □

Least Upper Bound Axiom   If a non-empty set has an upper bound, it must have a least upper bound.

The least upper bound axiom is sometimes called the supremum principle or the completeness axiom. Along with the axioms needed to show that the set of real numbers is an ordered field, this additional axiom enables us to develop careful proofs of the theorems in analysis for a function of one

real variable. The concepts of a least upper bound and a greatest lower
bound (see Definition 2) are generalizations for infinite sets of the concepts
of a maximum and minimum for a finite set. For example, if S is the set
$\{\frac{1}{n} : n \in \mathbf{N}\}$, then max $S = 1$, while min S is not defined because there
is no smallest element in S. However, sup S and inf S can be defined for
all sets if we permit the use of $\pm\infty$. The *extended real number system* is
the set $\mathbf{R} \cup \{\pm\infty\}$, and is a useful concept in a setting such as this one.

Definition 2   The number c is a *lower bound* for the set S if $c \leq x$
holds for all $x \in S$. The number g is a *greatest lower bound* for the set S
if g is a lower bound for S and $c \leq g$ holds for all lower bounds c of S. A
set S is *bounded* if it has both an upper bound and a lower bound.

It is also true that if a set S has a greatest lower bound, then that
value is unique, so we can speak of *the* greatest lower bound for S, and
denote it by glb S. The alternate notation is inf S, where inf denotes
infimum. If a non-empty set has no lower bounds, we express this by
glb S = inf S = $-\infty$.

Example 2   For the set $\mathbf{N}$ of natural numbers, inf $\mathbf{N} = 1$ and sup $\mathbf{N} =$
$+\infty$.   □

Example 3   Let S be the set determined by the sequence $\{x_n\}$ where
$x_n = \frac{n+2}{3n-1}$, namely $S = \{\frac{n+2}{3n-1} : n \in \mathbf{N}\}$. The first few elements of S are
$\frac{3}{2}, \frac{4}{5}, \frac{5}{8}$, and $\frac{6}{11}$, and we know that the elements approach $\frac{1}{3}$ from above
for large values of n. Therefore inf $S = \frac{1}{3}$ and sup $S = \frac{3}{2}$.   □

We discuss three main proof forms in Appendix 3. The direct proof is
considered the most desirable form and we try to use it whenever possible.
All the proofs in Section 4 were of this type. However, in trying to prove the
statement $p \to q$, we sometimes are unable to find any useful implications
of statement p. In such cases, we see if there are any useful implications
of $\neg q$. If we can construct a direct proof of $\neg q \to \neg p$, we say that we
have also proved $p \to q$ by a contrapositive proof. We use this technique
to prove the results in Lemma 5.1.

LEMMA 5.1   Let $a, b \in \mathbf{R}$.

(1) If $a \leq b + \epsilon$ for all $\epsilon > 0$, then $a \leq b$

(2) If $0 \leq a - b < \epsilon$ for all $\epsilon > 0$, then $a = b$

Proof of (1).

We first write (1) in the form

      p: $a \leq b + \epsilon$ for all $\epsilon > 0$

      q: $a \leq b$

and then in the contrapositive form

      ¬q: $a > b$

      ¬p: There is some $\epsilon_o > 0$, so that $a > b + \epsilon_o$

If we define $\epsilon_o = \frac{1}{2}(a - b)$, we can develop a simple proof of $\neg q \rightarrow \neg p$, and therefore a contrapositive proof of $p \rightarrow q$.   ■

The proof of (2) is left as an exercise.

The next lemma gives an equivalent form of Definitions 1 and 2, which will be useful for many proofs. It expresses the intuitive idea that there can be no "gaps" between the set S and the numbers sup S and inf S.

LEMMA 5.2   An upper bound u for the set S is the sup S if and only if for every $\epsilon > 0$, there is some $x_\epsilon \in S$, so that $u - \epsilon < x_\epsilon \leq u$. Similarly, a lower bound g for the set S is the inf S if and only if for every $\epsilon > 0$, there is some x_ϵ in S, so that $g \leq x_\epsilon < g + \epsilon$.

Comment.   We prove this result for sup S and leave the proof for inf S as an exercise. In order to help you recognize the difference between a direct proof and a contrapositive proof, we will restate this lemma more carefully and show all the steps in the proof. We assume first that u is an upper bound for the set S.

Proof of ⇐:

      p: For every $\epsilon > 0$, there is some $x_\epsilon \in S$ so that $u - \epsilon < x_\epsilon \leq u$.

      q: $u = \sup S$

<u>A Direct Proof</u>

    Let b be any upper bound for S.

    Let ϵ be any positive real number.

    There is some $x_\epsilon \in S$, so that $u - \epsilon < x_\epsilon \leq u$.

Also $x_\epsilon \leq b$.

Therefore $u - \epsilon < b$ or $u < b + \epsilon$.

Then $u \leq b$ by Lemma 5.1, and so $u = \sup S$ by Definition 1. ■

Proof of $\Rightarrow$:

p: $u = \sup S$

q: For every $\epsilon > 0$, there is some $x_\epsilon \in S$, so that $u - \epsilon < x_\epsilon \leq u$.

Restated in contrapositive form

$\neg q$: There is some $\epsilon_o > 0$, so that for each $x \in S$, either $x > u$ or $u - \epsilon_o \geq x$.

$\neg p$: $u \neq \sup S$

If there is some $x \in S$ for which $x > u$, then $u \neq \sup S$, because u is not even an upper bound for S. And if $x \leq u - \epsilon_o$ for all $x \in S$, then $u - \epsilon_o$ is an upper bound for S, and since $u - \epsilon_o < u$, it follows that $u \neq \sup S$. ■

LEMMA 5.3   Let S be any non-empty set.

(1) If $x \leq t$ for all $x \in S$, then $\sup S \leq t$

(2) If $x \geq t$ for all $x \in S$, then $\inf S \geq t$

Proof of (1).   By the Trichotomy law (see Appendix 2), if $\sup S \leq t$ did not hold, then $t < \sup S$ must hold. By Lemma 5.2, there must be some $x \in S$ so that $t < x < \sup S$, which is the negation of the hypothesis that $x \leq t$ for all $x \in S$. Therefore this proof is finished.   ■

You should identify the proof form used for each proof in this section; for example, which proof form was just used in this proof of Lemma 5.3?

Proof of (2).   This is left as an exercise.

The following lemma contains results that will be helpful in working with properties of the definite integral in Chapter 6.

LEMMA 5.4   Let the function f be defined on the set B and use the symbols $\inf_B f$ to represent $\inf\{f(x) : x \in B\}$ and $\sup_B f$ to represent $\sup\{f(x) : x \in B\}$.

(1) If $A \subset B$, then $\inf B \leq \inf A \leq \sup A \leq \sup B$

(2) If $A \subset B$, then $\inf_B f \leq \inf_A f \leq \sup_A f \leq \sup_B f$

(3) If g is defined on the set B, then

$$\inf_{B} f + \inf_{B} g \leq \inf_{B}(f+g) \leq \sup_{B}(f+g) \leq \sup_{B} f + \sup_{B} g$$

Proof of (1) and (2). These are left as an exercise.

Proof of (3). Let $\epsilon > 0$ be given. By Lemma 5.1, there is some $x_\epsilon \in B$ so that $(f+g)(x_\epsilon) \leq \inf_{B}(f+g) + \epsilon$, and therefore

$$\inf_{B} f + \inf_{B} g \leq f(x_\epsilon) + g(x_\epsilon) \leq \inf_{B}(f+g) + \epsilon$$

Since this holds for all $\epsilon > 0$, we have $\inf_{B} f + \inf_{B} g \leq \inf_{B}(f+g)$ by Lemma 5.1. Proofs for the remaining inequalities in (3) are left as an exercise. ■

The example of $f(x) = x$ and $g(x) = 1 - x$ on the interval $B = [0, 1]$ shows that strict inequality is possible in Lemma 5.4.

Comment. There are a number of results that can be proven from the least upper bound axiom. In fact, most of these results are equivalent to this axiom, which means that any of them could be chosen as the starting axiom. We will only prove enough results to obtain Theorem 7.1, the Bolzano-Weierstrass Theorem for sequences, which is our main goal in this chapter. There is an alternate proof for Theorem 7.1 that does not require the use of Theorems 5.5 and 5.6. Information on this is provided in a comment after the proof of Theorem 7.1.

Definition 3 We use $\{I_n\}$ to represent a sequence of closed intervals, where $I_n = [a_n, b_n]$. We call $\ell(I_n) = b_n - a_n$ the *length* of I_n. This sequence will be called *decreasing* if $I_{n+1} \subset I_n$ for all n. This also implies that $a_n \leq a_{n+1} < b_{n+1} \leq b_n$. A sequence of open intervals, or half-open intervals, can be defined in a similar way.

Definition 4 A *nest of closed intervals* is a decreasing sequence of closed intervals whose length goes to zero. Stated differently, if we let $I_n = [a_n, b_n]$, then $I_{n+1} \subset I_n$ for all n, and $\lim_{n \to \infty}(b_n - a_n) = 0$. A nest of open intervals can be defined in a similar way.

Example 4   The sequence $\{I_n\}$ where $I_n = [-\frac{1}{n}, 1+\frac{1}{n}]$ is a decreasing sequence of closed intervals, but it is not a nest because the length of I_n does not go to zero.   □

Example 5   The sequence $\{I_n\}$ where $I_n = (1-\frac{1}{n}, 1+\frac{1}{n})$ is a nest of open intervals because the sequence of open intervals is decreasing and also $\lim (b_n - a_n) = \lim \frac{2}{n} = 0$.   □

Example 6   The sequence $\{I_n\}$ where $I_n = (1, 1+\frac{1}{n})$ is also a nest of open intervals. There is one difference between Examples 5 and 6, which you can see by comparing the set $\cap_{n=1}^{\infty} I_n$ for these two examples.   □

Example 7   The sequence $\{I_n\}$ where $I_n = [1-\frac{1}{n}, 1+\frac{1}{n}]$ is a nest of closed intervals.   □

THEOREM 5.5 (The Nested Interval Property)

If $\{I_n\}$ is a nest of closed intervals, then there is exactly one value x that belongs to every I_n.

Proof.   If we let $I_n = [a_n, b_n]$, we can form the set $S = \{a_n : n \in \mathbf{N}\}$. This set S is non-empty, and bounded above since $a_n < b_1$ for all n. Therefore S must have a least upper bound, which we denote by x. We next show that x must be in every I_n.

First, $a_n \leq x$ since x is an upper bound for S. Also, $x \leq b_n$ must hold for all n, since if $x > b_{n_o}$ for some n_o, there would have to be some $n_1 > n_o$, so that $x > a_{n_1} > b_{n_o}$. This follows from Lemma 5.2. Since $b_{n_1} > a_{n_1}$, we have $b_{n_1} > b_{n_o}$, which is a contradiction to the assumption that $\{I_n\}$ is a nest.

To prove the uniqueness of x, suppose that the element y is in every I_n. For each $\epsilon > 0$, there is some N so that $\ell(I_n) < \epsilon$ for $n > N$. But then, $0 \leq |x - y| \leq \ell(I_n) < \epsilon$ implies by Lemma 5.1 that $x = y$, and so there is only one element in the set $\cap_{n=1}^{\infty} I_n$.   ∎

Definition 5   A real number x is called a *cluster point* (or an *accumulation point*) for the set S if for every $\epsilon > 0$, there is some $s_\epsilon \in S$ so that $s_\epsilon \neq x$ and $x - \epsilon < s_\epsilon < x + \epsilon$.

Example 8   The set of natural numbers **N** has no cluster points. For example, if $n \in$ **N**, the interval $(n - \frac{1}{2}, n + \frac{1}{2})$ contains no point of **N** different from n. A similar argument can be given for any $x \in$ **R/N**. Recall from Appendix 1 that **R/N** denotes the set of all real numbers that are not natural numbers.   □

Example 9   Let I be the interval (a, b). Every point $c \in I$ is a cluster point for I. For if $\epsilon > 0$ is given, and $\epsilon < \min\{c-a, b-c\}$, then $s_\epsilon = c + \frac{\epsilon}{2}$ meets the conditions of a cluster point as given in Definition 5.   □

THEOREM 5.6 (Bolzano-Weierstrass Theorem for Sets)
Every bounded, infinite set S of real numbers must have a cluster point.

Proof.   Since S is bounded, there must be an interval $I = [a, b]$, so that $S \subset I$. If I is bisected, then either the left half or the right half of I must contain an infinite number of values from S. Let $I_1 = [a, \frac{a+b}{2}]$ if $I_1 \cap S$ is infinite. Otherwise, let $I_1 = [\frac{a+b}{2}, b]$. Define I_2 as the left half of I_1 if this half contains an infinite number of terms of S, and as the right half if there are only a finite number of terms of S in the left half of I_1. Continuing in this way, we can inductively define a nest of closed intervals $\{I_n\}$ with the property that $\ell(I_n) = \frac{b-a}{2^n}$ and $I_n \cap S$ is infinite. By the nested interval property, there is a unique value $x \in \cap I_n$, and we claim that x must be a cluster point for S.

  To prove this, let $\epsilon > 0$ be given. Since $\ell(I_n)$ approaches 0, there must be some N so that $I_n \subset (x - \epsilon, x + \epsilon)$ for all $n \geq N$. Since $I_N \cap S$ is infinite, there must be some (in fact, infinitely many) $s \in S$, so that $s \neq x$, and s is in the interval $(x - \epsilon, x + \epsilon)$. This verifies that x is a cluster point for S. ∎

Example 10   The set $S = \{\sin \frac{n\pi}{2} : n \in$ **N**$\}$ is a bounded set, but it does not have any cluster points because it is a finite set, consisting of the three elements $\{0, +1, -1\}$.   □

Example 11   The set $S = \{\sin n : n \in$ **N**$\}$ is a bounded set with an infinite number of elements, so it must have at least one cluster point. However, Theorem 5.6 gives no method for finding such a cluster point, or for deciding how many cluster points this set may have.   □

  It is possible to show that the LUB axiom, the nested interval property, and the Bolzano-Weierstrass Theorem for sets are all equivalent. In

addition, there are other results such as the Dedekind cut axiom of continuity, the Heine-Borel Theorem, and the Bolzano-Weierstrass Theorem for
sequences, which are also equivalent to the LUB axiom. Thus any one of
these six results could be chosen as the starting axiom, with the other five
being theorems that could be derived from the chosen axiom, and there are
many different orders in which to do this. We follow the traditional mathematical approach of choosing the statement that is conceptually simplest
for the beginning axiom.

The Heine-Borel Theorem is stated and proved in Section 13, the
Bolzano-Weierstrass Theorem for sequences is proved in Section 7, and
a discussion of the Dedekind cut axiom may be found in a book such as
Principles of Mathematical Analysis by Walter Rudin.

There are some other results that are consequences of the LUB axiom
without being equivalent to it. Three such results are listed below, and
will be used in several places throughout this book.

LEMMA 5.7 (Archimedean Axiom - Form 1)

For any positive real numbers a and b, there is a natural number n so that
$na > b$.

Proof.   This is an example of a proof by contradiction. We begin by
assuming the statement $\neg q$: $na \leq b$ for all $n \in \mathbf{N}$. If we apply the
LUB axiom to the set $S = \{na : n \in \mathbf{N}\}$, there must be a least upper
bound for S, which we can denote by u. By Lemma 5.2, using $\epsilon = a$,
there is some $n_o \in \mathbf{N}$ so that $u - a < n_o a \leq u$. But then $(n_o + 1)a \in S$
and $(n_o + 1)a > u$, which implies that $u \neq \sup S$. We have therefore
completed the proof by contradiction of the statement r : $u = \sup S$.   ∎

COROLLARY 5.8 (Archimedean Axiom - Form 2)

For any real number $t > 0$, there is a natural number n so that $0 < \frac{1}{n} < t$.
Proof.   This is left as an exercise.

COROLLARY 5.9 (Density Theorem)

For any two real numbers $x < y$, there is a rational number q so that
$x < q < y$.

Proof.   Assume first that both numbers are positive, so we can write
$0 < x < y$. Then by Corollary 5.8, there is $n \in \mathbf{N}$, so that $\frac{1}{n} < y - x$.
This is equivalent to $nx < ny - 1$. There is also a natural number m so that
$m - 1 \leq nx < m$, from which we have $x < \frac{m}{n}$. Finally, $m - 1 < ny - 1$,

which yields $\frac{m}{n} < y$. We can therefore let $\frac{m}{n}$ be the desired rational number q, since it is defined as the quotient of two integers.   ∎

The other two cases, when both x and y are negative, and when one is negative and one is positive, are left as an exercise.

Example 12   Let F be the set of rational numbers in the interval $[0, 1]$. Then every real number in $[0, 1]$ is a cluster point for F. For example, $\frac{1}{2}$ is a cluster point for F, since if (a, b) is an interval containing $\frac{1}{2}$ so that $0 < a < \frac{1}{2} < b < 1$, then by the density theorem, there is a rational number r so that $a < r < \frac{1}{2}$. This verifies that $\frac{1}{2}$ is a cluster point for F. A similar argument could be used for any real number in $[0.1]$.   □

EXERCISES 5

1. Find the sup S and inf S for the following sets.

    (a) S is the set $\mathbf{Z}$ of integers.

    (b) $S = \{2 + \frac{(-1)^n}{n} : n \in \mathbf{N}\}$

    (c) $S = \{\frac{n-1}{2n+1} : n \in \mathbf{N}\}$

    (d) $S = $ set of rational numbers r so that $0 < r^2 < 2$.

2. Give a statement of the Greatest Lower Bound axiom, referring to the LUB axiom as an example.

3. Prove that inf S is uniquely defined for a set S.

4. Prove part (2) of Lemma 5.1.

5. It is usually possible, although not necessarily desirable, to replace a direct proof by a contrapositive proof. Illustrate this for the direct proof given for part (1) of Lemma 5.2.

6. Prove the second half of Lemma 5.2.

7. Prove part (2) of Lemma 5.3.

8. Prove parts (1) and (2) of Lemma 5.4.

9. Finish the proof of result (2) in Lemma 5.4.

10. For a given set S, define the set $-S = \{-x : x \in S\}$. Prove that $\sup(-S) = -\inf S$ and $\inf(-S) = -\sup S$.

11. Prove that the Greatest Lower Bound (GLB) axiom can be obtained as a consequence of the LUB axiom. This means that you should begin with a set S that is bounded below (suppose $B \leq x$ for all $x \in S$). The next step is to obtain a set that is bounded above, so that the LUB axiom can be applied.

12. Which of the following sequences are a nest of closed intervals? Give reasons.
    (a) $I_n = [-\frac{1}{n}, 0]$
    (b) $I_n = [(-1)^n, 1]$
    (c) $I_n = [n, n+1]$
    (d) $I_n = [\sin \frac{1}{n}, \frac{1}{n}]$. Note that for this to be a well-defined interval, we need the fact that $\sin x < x$ for $x \leq 1$.

13. Find the cluster points for the following sets.
    (a) $S = \{\frac{1}{n} : n \in \mathbf{N}\}$
    (b) $S = \{1, 2, 3, \ldots, 10\}$
    (c) $S = [0, 1) = \{x \in \mathbf{R} : 0 \leq x < 1\}$

14. Use Lemma 5.7 to prove Corollary 5.8.

15. Prove the remaining two cases of Corollary 5.9.

16. What is the difference between the set $\cap I_n$ in Example 5 and the set $\cap I_n$ in Example 6?

17. In addition to the proofs provided in this section, what additional result is needed in order to prove that the LUB axiom, the nested interval property, and the Bolzano-Weierstrass Theorem for sequences are all equivalent?

18. Consider the set $S = \{\tan n : n \in \mathbf{N}\}$. What can be said about the set of cluster points for S?

19. Let x be the unique element in the nest of closed intervals in Theorem 5.5. For any $\epsilon > 0$, prove there is some $n \in \mathbf{N}$, so that $I_n \subset (x - \epsilon, x + \epsilon)$.

SECTION 6 MONOTONE AND BOUNDED SEQUENCES

We are now able to prove the first of two important results about bounded sequences, namely that if a bounded sequence is monotone, then it must converge. In Section 8 we will prove the second result, which is that every bounded sequence must have a convergent subsequence.

Definition 1   The sequence $\{x_n\}$ is *increasing* if $x_n < x_{n+1}$ holds for all n. The sequence $\{x_n\}$ is *decreasing* if $x_{n+1} < x_n$ holds for all n. The sequence $\{x_n\}$ is *nondecreasing* if $x_n \leq x_{n+1}$ or *nonincreasing* if $x_{n+1} \leq x_n$ holds for all n. The sequence $\{x_n\}$ is *monotone* if it satisfies any of the above conditions. The sequence $\{x_n\}$ is *eventually monotone* if there is some $k \geq 1$ so that one of the inequalities above is satisfied for all $n > k$.

Example 1   The sequence $\{1 - \frac{1}{n}\}$ is increasing and therefore also nondecreasing. The sequence $\{\frac{\ln n}{n}\}$ is not decreasing, but is eventually decreasing because $\frac{\ln(n+1)}{n+1} \leq \frac{\ln n}{n}$ holds for $n \geq 4$. The sequence $\{\sin n\}$ is not monotone.   □

THEOREM 6.1 (Monotone and bounded sequences)
If $\{x_n\}$ is a sequence that is monotone and bounded, then $\{x_n\}$ converges.

Proof.   If the sequence $\{x_n\}$ is eventually nondecreasing and bounded above, then there is some $B > 0$ and $k \geq 1$, so that $x_n \leq x_{n+1} \leq B$ holds for all $n \geq k$. Since the set

$$S = \{x_n : n \in \mathbf{N} \text{ and } n \geq k\}$$

has an upper bound, it must have a least upper bound u, by the LUB axiom. We next assert that the sequence $\{x_n\}$ converges to u. This follows from the definition of u, since for any $\epsilon > 0$, there is some $N > k$, so that $u - \epsilon < x_N \leq u$. Then by Exercise 9, $x_N \leq x_n$ for $n > N$, and therefore $u - \epsilon < x_N \leq x_n \leq u$.   ■

It is left as an exercise to prove that a sequence that is eventually nonincreasing and bounded below must also converge.

Comment.   An important goal in this chapter is to introduce you to a variety of the proof forms used in analysis. You probably recall the typical geometry proofs from high school, which were in the form of a series of steps, each with a supporting reason. Our proof of Theorem 4.1 resembled that form. Since then, most proofs have been presented in a paragraph format, which is the more familiar style found in advanced mathematics. The first proofs you do in this course will likely be in the step-reason format, but gradually the transition should be made to the paragraph format. For comparison purposes, a step-reason format of the proof of Theorem 6.1 is now presented. You should supply the reasons, as well as any missing steps.

(1) There is $B > 0$ and $k \geq 1$ so that $x_n \leq x_{n+1} \leq B$ for all $n \geq k$.

(2) Define the set $S = \{x_n : n \in \mathrm{N} \text{ and } n \geq k\}$.

(3) There is a value $u = \sup S$.

(4) Let $\epsilon > 0$ be given.

(5) There is some $N \geq k$ so that $x_N \in S$ and $u - \epsilon < x_N \leq u$.

(6) For all $n > N$, $u - \epsilon < x_N \leq x_n \leq u$.

(7) Therefore $\{x_n\}$ converges to u. ∎

An important use of Theorem 6.1 is the proofs of the convergence tests for series of positive terms. These are known by such familiar names as the Comparison Test, the Ratio Test, and the Integral Test. They will be discussed and proven in Chapter 8. Theorem 6.1 may also be used to establish the convergence of some sequences, such as the sequence $\{x_n\}$, where $x_n = 1 + \frac{1}{2} + \cdots + \frac{1}{n} - \ln n$, which was discussed in Example 2.10. Once we have shown that a sequence converges by Theorem 6.1, we can try to use another method to find the limit of the sequence.

We next look at some recursively defined sequences, which furnish another setting for the use of Theorem 6.1. These sequences also provide an opportunity to use the idea of mathematical induction. A sequence is said to be *recursively defined* if the term x_n is defined in terms of one or more preceding terms, rather than by an explicit formula.

Example 2   Let the sequence $\{x_n\}$ be recursively defined by the formula $x_{n+1} = \left(1 - \frac{1}{(n+1)^2}\right) x_n$, where $x_1 = 1$. Show that $\{x_n\}$ converges to $\frac{1}{2}$.

Solution.   The first few terms are $x_2 = \frac{3}{4}$, $x_3 = \frac{2}{3}$, and $x_4 = \frac{5}{8}$, which suggests that the sequence $\{x_n\}$ may be decreasing and bounded below. This can easily be established. Since $1 - \frac{1}{(n+1)^2} > 0$, all the x_n terms are positive, and so 0 is a lower bound. Since $1 - \frac{1}{(n+1)^2} < 1$, we have $x_{n+1} < x_n$, and so the sequence is decreasing. As a result, we know that $\{x_n\}$ must converge by Theorem 6.1, and we next seek to find the value for $\lim_{n \to \infty} x_n$. We can do this by the use of mathematical induction.

Looking at the first few terms above, we are led to conjecture that $x_n = \frac{n+1}{2n}$. If this were true, we would have an explicit formula for x_n, and would be able to conclude that $\{x_n\}$ converges to $\frac{1}{2}$. We complete this example with a proof that the proposition

$$P(n) \ : \ x_n \ = \ \frac{n+1}{2n}$$

is true for all n. We know that it is true for $n = 1$, 2, 3, and 4 from our calculations above. We need to show that $P(n+1) : x_{n+1} = \frac{n+2}{2n+2}$ follows from the induction assumption that $P(n)$ is true. Referring to the definition of x_{n+1} above,

$$x_{n+1} \ = \ \left(1 - \frac{1}{(n+1)^2}\right) x_n$$

$$= \ \left(1 - \frac{1}{(n+1)^2}\right)\left(\frac{n+1}{2n}\right)$$

$$= \ \frac{n+2}{2n+2}$$

and the solution is complete.   □

Example 3   Let $\{x_n\}$ be recursively defined by the formula

$$x_{n+1} \ = \ \frac{1}{2}\left(x_n + \frac{2}{x_n}\right)$$

where x_1 is a positive real number. Show that $\{x_n\}$ converges.

Solution.   If we start with the value $x_1 = 1$, we have $x_2 = \frac{3}{2}$ and $x_3 = \frac{17}{12}$. If we start with the value $x_1 = 4$, we have $x_2 = \frac{9}{4}$ and $x_3 = \frac{113}{72}$. Although the terms of the sequence vary with the choice of x_1, the limit of the sequence is not affected by this choice. If we knew that the sequence converged to a value x, then $\lim_{n\to\infty} x_n = x$ and also $\lim_{n\to\infty} x_{n+1} = x$ (see Exercise 8). This implies that

$$\lim_{n\to\infty} x_{n+1} = \frac{1}{2}\left(\lim_{n\to\infty} x_n + \frac{2}{\lim_{n\to\infty} x_n}\right)$$

which can be rewritten as $x = \frac{1}{2}(x + \frac{2}{x})$, and then simplified to $x^2 = 2$. At this point, we know that if the sequence converges, it must converge to $\sqrt{2}$.

Before the assertion that $\{x_n\}$ converges to $\sqrt{2}$ can be made, we must use Theorem 6.1 to show that the sequence does indeed converge. We first notice from the definition that $x_{n+1} > 0$ for all n, and so the sequence is bounded below. Then we observe that $x_{n+1} = \frac{1}{2}(x_n + \frac{2}{x_n})$ is equivalent to $x_n{}^2 - 2x_{n+1}x_n + 2 = 0$, which in turn is equivalent to

$$(x_n - x_{n+1})^2 + (2 - (x_{n+1})^2) = 0$$

This implies that $2 - (x_{n+1})^2 \leq 0$, so that $x_n{}^2 - 2 \geq 0$ for all n, and therefore,

$$x_n - x_{n+1} = x_n - \frac{1}{2}x_n - \frac{1}{x_n} = \frac{x_n^2 - 2}{2x_n} \geq 0$$

so $\{x_n\}$ is nonincreasing, and therefore convergent. You may recognize that the formula above is the same as the one used in calculus to approximate $\sqrt{2}$ by Newton's method. □

Example 4   Show that the sequence $\{x_n\}$ diverges, when $x_1 = 1$ and $x_{n+1} = 2x_n + 1$.

Solution.   The purpose of this example is to show that the technique used in Example 3 to find the limit of a recursively defined sequence cannot be safely used, unless it is first shown that the sequence converges. For if we let $x = \lim_{n\to\infty} x_n = \lim_{n\to\infty} x_{n+1}$, we obtain the equation $x = 2x + 1$, or $x = -1$. But this is meaningless, since this sequence is $\{1, 3, 7, 15, \ldots\}$, which diverges to $+\infty$. □

Example 5   Apply Theorem 6.1 to the sequence $\{x_n\}$, which is recursively defined by $x_1 = 1$, $x_2 = 2$, and $x_n = \frac{1}{2}(x_{n-1} + x_{n-2})$ for $n > 2$.

Solution.   By writing out the first several terms, we observe that the sequence is bounded, but not monotone. Therefore, Theorem 6.1 cannot be used for this example. We will return to this example in Section 7 and use a result about subsequences there to prove that this sequence converges. You may have noticed that without the factor of $\frac{1}{2}$, this sequence would be the well-known Fibonacci sequence $x_n = x_{n-1} + x_{n-2}$.   □

In the next example, we use the *binomial theorem expansion*. This important formula is used at several places in an analysis course, and can be stated as

$$(a+b)^n = \sum_{i=0}^{n} \binom{n}{i} a^{n-i} b^i$$

where $\binom{n}{i}$ is the symbol for the number of combinations of n things taken i at a time, and can be evaluated by the formula

$$\binom{n}{i} = \frac{n!}{i!(n-i)!}$$

This formula is developed and widely used in a course in discrete mathematics or probability. To illustrate,

$$(a+b)^6 = \sum_{i=0}^{6} \binom{6}{i} a^{6-i} b^i$$

$$= \binom{6}{0} a^6 b^0 + \binom{6}{1} a^5 b^1 + \cdots + \binom{6}{6} a^0 b^6$$

$$= a^6 + 6a^5 b + 15a^4 b^2 + 20a^3 b^3 + 15a^2 b^4 + 6ab^5 + b^6$$

Example 6   Apply Theorem 6.1 to the sequence $\{e_n\}$, which is explicitly defined by $e_n = \left(1 + \frac{1}{n}\right)^n$.

Solution.   We are able to show that this sequence is monotone increasing and bounded above, so it must converge. In fact, this sequence converges to the well-known value of $e \approx 2.71828$. The first time that the letter e was used in print to represent this value was in Euler's *Mechanica* in 1733 on page 95. Prior to this time, Euler used the letter c for this value.

Using the binomial theorem, we write

$$\left(1 + \frac{1}{n}\right)^n = \binom{n}{0} 1^n \left(\frac{1}{n}\right)^0 + \binom{n}{1} 1^{n-1} \left(\frac{1}{n}\right)^1 + \binom{n}{2} 1^{n-2} \left(\frac{1}{n}\right)^2 + \cdots$$

$$\cdots + \binom{n}{n} 1^0 \left(\frac{1}{n}\right)^n$$

$$= 1 + n \frac{1}{n} + \frac{n(n-1)}{2} \frac{1}{n^2} + \frac{n(n-1)(n-2)}{3!} \frac{1}{n^3} + \cdots + \frac{1}{n^n}$$

$$= 1 + 1 + \frac{1}{2!}\left(1 - \frac{1}{n}\right) + \frac{1}{3!}\left(1 - \frac{1}{n}\right)\left(1 - \frac{2}{n}\right) + \cdots$$

$$\cdots + \frac{1}{n!}\left(1 - \frac{1}{n}\right)\left(1 - \frac{2}{n}\right) \cdots \left(1 - \frac{n-1}{n}\right)$$

$$< 1 + 1 + \frac{1}{2!} + \frac{1}{3!} + \cdots + \frac{1}{n!}$$

It is left as an exercise to show that

$$\frac{1}{n^n} = \frac{1}{n!}\left(1 - \frac{1}{n}\right)\left(1 - \frac{2}{n}\right) \cdots \left(1 - \frac{n-1}{n}\right)$$

So by the result in Example 2.1 that $2^n \leq n!$ or $\frac{1}{n!} \leq \frac{1}{2^n}$ for $n \geq 4$, we have

$$e_n \leq 1 + 1 + \frac{1}{2!} + \frac{1}{3!} + \frac{1}{2^4} + \frac{1}{2^5} + \cdots + \frac{1}{2^n}$$

$$= 2\frac{2}{3} + \frac{1}{16}\left(1 + \frac{1}{2} + \frac{1}{4} + \cdots + \frac{1}{2^{n-4}}\right)$$

$$= 2\frac{2}{3} + \frac{1}{16}\left(2 - \frac{1}{2^{n-4}}\right)$$

$$< 2\frac{2}{3} + \frac{1}{8} < 3$$

and so $\{e_n\}$ is bounded above.

By a similar approach, we establish in the exercises that $e_{n+1} \geq e_n$, and therefore $\{e_n\}$ will converge by Theorem 6.1. By an approximation process, we are able to show that $\lim_{n \to \infty} e_n \approx 2.71828$, and we denote this important limit value by the letter e. □

EXERCISES 6

1. Observe that the recursively defined sequence in Example 2 is the same as the explicitly defined sequence in Exercise 4.2.

2. In Example 3, we showed that $\{x_n\}$ was nonincreasing, yet this doesn't seem to be the case since $x_1 = 1$ and $x_2 = \frac{3}{2}$. Where is the problem?

3. Write out the first 8 terms for the sequence in Example 5. Then observe that the subsequence (this concept will be formally defined in Section 7) of even subscripted terms, namely, $\{x_2, x_4, x_6, x_8, \ldots\}$ appears to be decreasing, while the subsequence of odd subscripted terms, namely $\{x_1, x_3, x_5, x_7, \ldots\}$ appears to be increasing. Do you see a pattern so that you could write an explicit formula for x_{2n}? This problem will be reconsidered in Sections 7 and 8.

4. Let $x_n = 1 + \frac{1}{4} + \frac{1}{9} + \cdots + \frac{1}{n^2}$.

   (a) Prove that $\{x_n\}$ is increasing.

   (b) Prove that $\{x_n\}$ is bounded above. Hint: Use the fact that

   $$\frac{1}{i^2} \leq \frac{1}{i(i-1)} = \frac{1}{i-1} - \frac{1}{i}.$$

5. Let $x_1 = 1$ and $x_{n+1} = \sqrt{2 + x_n}$.

   (a) Use mathematical induction and Theorem 6.1 to prove that $\{x_n\}$ converges.
   (b) Find $\lim_{n \to \infty} x_n$.

6. Let $x_1 = a > 0$ and $x_{n+1} = x_n + \frac{1}{x_n}$ for $n > 1$. Prove that $\{x_n\}$ must diverge.

7. Prove that the sequence $\{x_n\}$ converges, where $x_n = \frac{1}{n+1} + \frac{1}{n+2} + \cdots + \frac{1}{2n}$. Write out the first several terms to see the pattern. In Exercise 24.9, you are asked to show that this sequence converges to $\ln 2$.

8. Prove that if $\lim_{n \to \infty} x_n = x$, then $\lim_{n \to \infty} x_{n+k} = x$, for any $k > 0$.

9. Prove that if $\{x_n\}$ is a nondecreasing sequence, then $x_m \leq x_n$ whenever $m < n$.

10. Let $x_{n+1} = \frac{1}{2}(x_n + \frac{a}{x_n})$ where $x_1 > 0$ and $a > 0$. Prove that $\{x_n\}$ converges to $\sqrt{a}$. Use Example 3 as a guide.

11. Prove Theorem 6.1 for the case when $\{x_n\}$ is nonincreasing and bounded below. It is instructive to do two distinct proofs.

    (a) Write a proof patterned after the form of the proof given in the text for a nondecreasing sequence.

    (b) Write a proof using the result proved for a nondecreasing sequence, instead of the method.

12. Write out the binomial theorem expansion for $(a + b)^7$.

13. In Example 6, show that $\{e_n\}$ is monotone increasing by first expanding e_{n+1} by the binomial theorem, and then showing that it is greater than the expression for e_n that was obtained in Example 6.

14. As part of the work in Example 6, show that

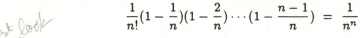

$$\frac{1}{n!}(1 - \frac{1}{n})(1 - \frac{2}{n})\cdots(1 - \frac{n-1}{n}) = \frac{1}{n^n}$$

Just look

15. Make a list of the various methods that we have to find the limit of a convergent sequence. You may include methods from calculus, even if we have not discussed them yet. You may add to this list as we proceed further in this course.

SECTION 7 SUBSEQUENCES

We come now to the most important result in this chapter, in the sense that it will be used to prove the results about continuous functions in Chapter 4 that are needed to prove the Mean Value Theorem in Chapter 5 and ultimately, the Fundamental Theorem of Calculus in Chapter 6. We refer to the Bolzano-Weierstrass Theorem, which is stated here in the terminology of sequences. We will use the theorems from Section 5 in our work. We need first to provide an introduction to the concept of subsequences and will also derive some practical results involving this idea.

Definition 1   A *subsequence* of a given sequence is itself a sequence obtained by choosing an infinite number of the terms of the original sequence in the order in which they occur. If $\{x_n\}$ denotes the original sequence, then $\{x_{n_k}\}$ denotes a subsequence, where the subscript $\{n_k\}$ is itself a sequence with the variable k, so that $n_1 < n_2 < \cdots < n_k < \cdots$

Comment.   The subsequence $\{x_{n_k}\}$ can also be considered as the composition of the functions $x(n)$ and $n(k)$.

Example 1   If $x_n = x(n) = \frac{1}{n}$ and $n_k = n(k) = 2k$, then $\{x_{n_k}\} = \{x_{2k}\} = \frac{1}{2k}$ is the subsequence of even subscripted terms of the sequence $\{x_n\}$. $\square$

Definition 2   The number x is called *a limit point* for the sequence $\{x_n\}$ if there is a subsequence that converges to x. An alternate way to state this is that every ϵ-interval about x contains infinitely many terms of the sequence $\{x_n\}$.

Example 2   Let $\{x_n\} = \{(-1)^n\}$ be the given sequence. Since the subsequence $\{x_{2k}\}$ converges to 1, and the subsequence $\{x_{2k-1}\}$ converges to -1, the sequence has two limit points. But the set $S = \{x_n : n \in \mathbf{N}\}$ consists of only two elements, and therefore has no cluster points. So there is a difference between a limit point of the sequence $\{x_n\}$, and a cluster point of the set $\{x_n : n \in \mathbf{N}\}$. $\square$

Example 3   The sequence $\{x_n\} = \{(-1)^n + \frac{1}{n}\}$ has the limit points of -1 and $+1$, which you should verify. The set $S = \{x_n : n \in \mathbf{N}\}$ also has ± 1 as cluster points. How does this differ from the set S in Example 2? $\square$

THEOREM 7.1 (Bolzano-Weierstrass Theorem for sequences)
A bounded sequence must have at least one convergent subsequence.

Proof.   Let $\{x_n\}$ be a bounded sequence. If the set $S = \{x_n : n \in \mathbf{N}\}$ is finite, then at least one sequence value must occur infinitely often, and a convergent subsequence can be determined, all of whose terms assume this constant value.

If the set S is infinite, then since it is assumed to be bounded, there is a cluster point x by Theorem 5.6. We can finish this proof by constructing a subsequence that converges to x.

For $\epsilon = 1$, by the definition of a cluster point for a set, there is an element $x_{n_1} \in S$, so that x_{n_1} is in the interval $(x - 1, x + 1)$.

For $\epsilon = \frac{1}{2}$, in a similar way, there is an element $x_{n_2} \in S$, so that $n_2 > n_1$ and $x_{n_2} \in (x - \frac{1}{2}, x + \frac{1}{2})$.

Continuing in this inductive way, we assume that we have defined the ordered n-tuple $\{x_{n_1}, x_{n_2}, \ldots, x_{n_k}\}$, so that $x_{n_i} \in (x - \frac{1}{i}, x + \frac{1}{i})$ and $n_j > n_i$ for $j > i$.

It is then possible to choose $x_{n_{k+1}} \in S$, where $n_{k+1} > n_k$, and so that $x_{n_{k+1}} \in (x - \frac{1}{k+1}, x + \frac{1}{k+1})$. We have therefore defined by induction a subsequence $\{x_{n_k}\}$ with the property that $x_{n_k} \in (x - \frac{1}{k}, x + \frac{1}{k})$. Therefore $\{x_{n_k}\}$ converges to x, since for every $\epsilon > 0$, there is some K so that $\frac{1}{K} < \epsilon$, and for which $|x_{n_k} - x| < \frac{1}{k} < \epsilon$ for all $k > K$.   ∎

Comment.   Now that we have constructed the above proof for the Bolzano-Weierstrass Theorem for sequences based on Theorem 5.5 (Nested Interval Property) and Theorem 5.6 (Bolzano-Weierstrass Theorem for sets), we mention an alternate approach. A recent article in the *American Mathematical Monthly*[1] contained a short proof that any bounded sequence has a monotone subsequence. Then Theorem 6.1 guarantees that this subsequence converges, and so Theorem 7.1 is quickly proven.

---

[1] "On monotone subsequences" by Donald Newman and T. D. Parsons, Jan 1988, pages 44-45.

Example 4   The sequence $\{x_n\} = \{\sin n\}$ is a bounded sequence, so it must have at least one convergent subsequence. $\square$

THEOREM 7.2   If a sequence converges to x, then every subsequence converges to x also.

Proof.   You should not let the simplicity of this proof be hidden by the symbolism. Let the sequence $\{x_n\}$ converge to x, and let $\{x_{n_k}\}$ be any subsequence of this sequence. Let $\epsilon > 0$ be given. In order to show that $\{x_{n_k}\}$ converges to x, we need to find some K so that $|x_{n_k} - x| < \epsilon$ for all $k > K$.

But there is some N so that $|x_n - x| < \epsilon$ for all $n > N$. Also there is some K so that $n_K > N$. Then if $k > K$ we have $n_k > n_K$ and so $|x_{n_k} - x| < \epsilon$.   ∎

Example 5   Find $\lim_{n \to \infty}(1 + \frac{1}{2n})^{2n}$.

Solution.   In Example 6.6, we proved the sequence $\{e_n\} = \{(1 + \frac{1}{n})^n\}$ converged to a value we denoted by e. The sequence in this example is a subsequence of $\{e_n\}$, since $e_{2n} = (1 + \frac{1}{2n})^{2n}$, and therefore the subsequence must converge to e also. $\square$

Theorem 7.2 has several practical applications, which are now listed without proof. The term *corollary* is used to refer to a result that is easily obtained as a consequence of a given theorem.

COROLLARY 7.3   If the sequence $\{x_n\}$ converges, and there is a subsequence that converges to x, then $\{x_n\}$ must converge to x.

COROLLARY 7.4   If there are two subsequences of $\{x_n\}$ that converge to different values, then $\{x_n\}$ must diverge.

COROLLARY 7.5   If a subsequence of $\{x_n\}$ diverges, then $\{x_n\}$ must also diverge.

COROLLARY 7.6   If there are two or more subsequences of $\{x_n\}$ that each converge to x, and all but a finite number of values of $\{x_n\}$ belong to at least one of the subsequences, then $\{x_n\}$ converges to x also.

Example 6   We use Corollary 7.4 to assert that $\{x_n\} = \{(-1)^n\}$ diverges, because the subsequence $\{x_{2n}\}$ converges to 1, while the subsequence $\{x_{2n-1}\}$ converges to -1. $\square$

Example 7   We can use Corollary 7.6 to show that the sequence $\{x_n\}$ of Example 6.4, where $x_n = \frac{1}{2}(x_{n-1} + x_{n-2})$, $x_1 = 1$, and $x_2 = 2$, converges to $\frac{5}{3}$.

Solution.   We first list several terms from the subsequence of even terms, and then try to find a pattern. We have $x_2 = 2$, $x_4 = \frac{7}{4}$, $x_6 = \frac{27}{16}$, and $x_8 = \frac{107}{64}$. Also, $x_4 - x_2 = -\frac{1}{4}$, $x_6 - x_4 = -\frac{1}{16}$, $x_8 - x_6 = -\frac{1}{64}$, and so on. This pattern is a consequence of the averaging nature of this definition.

Also, $x_8 = x_2 + (x_4 - x_2) + (x_6 - x_4) + (x_8 - x_6) = 2 - \frac{1}{4} - \frac{1}{16} - \frac{1}{64}$. In general, $x_{2n} = 2 - \frac{1}{4} - \frac{1}{16} - \cdots - \frac{1}{4^{n-1}}$. By the use of a geometric series result (see Section 9 if needed),

$$\lim x_{2n} = 2 - \frac{1}{4}\left(\sum_{n=0}^{\infty} \left(\frac{1}{4}\right)^n\right) = 2 - \frac{1}{4}\cdot\frac{1}{1-\frac{1}{4}} = 2 - \frac{1}{3} = \frac{5}{3}$$

It is left as an exercise for you to present a similar argument for the subsequence of odd terms, which begins with $x_1 = 1$, $x_3 = \frac{3}{2}$, $x_5 = \frac{13}{8}$, and $x_7 = \frac{53}{32}$. $\square$

Example 8   The sequence as defined below has an infinite number of limit points, because for every $n \in \mathbf{N}$, there is a subsequence that converges to n. On the other hand, the set $S = \{x_n : n \in \mathbf{N}\}$ has no cluster points.

$$\{x_n\} = \{1, 1, 2, 1, 2, 3, 1, 2, 3, 4, 1, 2, 3, 4, 5, \ldots\}   \square$$

We conclude this section with a brief introduction to the idea of summability. For a given sequence $\{x_n\}$, we define the *arithmetic mean sequence* $\{t_n\}$ where $t_n = \frac{1}{n}(x_1 + x_2 + \cdots + x_n)$. If $\{x_n\}$ converges to x, then we can prove that $\{t_n\}$ must also converge to x. However, there are some divergent sequences $\{x_n\}$ for which $\{t_n\}$ converges. Such sequences are called *summable*, or *Cesaro-summable*, after the individual who developed this concept.

Example 9   Let $\{x_n\} = \{1, 0, 1, 0, 1, 0, \ldots\}$. Then $t_1 = \frac{1}{1}$, $t_2 = \frac{1}{2}$, $t_3 = \frac{2}{3}$, $t_4 = \frac{2}{4}$, $t_5 = \frac{3}{5}$, and so on. In general, we have $t_{2n} = \frac{1}{2}$ and $t_{2n-1} = \frac{n}{2n-1}$. By Corollary 7.6, the arithmetic mean sequence $\{t_n\}$ converges to $\frac{1}{2}$, even though the sequence $\{x_n\}$ diverges by oscillation. $\square$

EXERCISES 7

1. Find all the limit points for the sequences $\{x_n\}$ as defined below.
   (a) $x_n = \sin\frac{n\pi}{2}$
   (b) $x_n = [\sin n]$ This is the greatest integer function.
   (c) $x_n = n(1 + (-1)^n)$
   (d) $x_n = \frac{n+1}{2n+3}$

2. Let the set S be defined as $\{x_n : n \in \mathbf{N}\}$ for each sequence in Exercise 1. Find all the cluster points for each of these sets, and compare with your answers in Exercise 1 for the limit points of the sequences.

3. Finish the work in Example 7 by showing that the subsequence $\{x_{2n-1}\}$ converges to $\frac{5}{3}$. Then apply Corollary 7.6 to assert that the original sequence $\{x_n\}$ in Example 7 also converges to $\frac{5}{3}$.

4. Let $x_1 = a$, $x_2 = b$, and $x_n = \frac{1}{2}(x_{n-1} + x_{n-2})$ for $n \geq 3$. Make a conjecture for the value of $\lim x_n$. You do not need to prove this result. Refer to Exercise 3 and Example 7.

5. What would you guess to be the collection of the limit points for the sequence $\{x_n\} = \{\sin n\}$ in Example 4? How is this question related, if at all, to the problem of finding the collection of limit points for the set S of rational numbers in the interval $[-1, 1]$?

6. Give an example of a sequence with 4 limit points.

7. Use Theorem 7.2 to obtain simple proofs of Corollaries 7.3 to 7.6.

8. Use Theorem 7.2 and $\lim_{n \to \infty}(1 + \frac{1}{n})^n = e$ to find the limits of the following sequences, if possible.
   (a) $\{(1 + \frac{1}{3n})^n\}$
   (b) $\{(1 - \frac{1}{n})^{3n}\}$
   (c) $\{(2 + \frac{1}{n})^n\}$
   (d) $\{(1 + \frac{3}{n})^n\}$ — Because this is not a subsequence of $\{(1 + \frac{1}{n})^n\}$, you can only give a partial answer to this problem.

9. Find the value of the arithmetic mean sequence $\{t_n\}$ for the following sequences.
   (a) $\{x_n\} = \{1, 0, 0, 1, 0, 0, 1, 0, 0, \ldots\}$

   (b) $\{x_n\} = \{1, -1, 1, -1, 1, -1, \ldots\}$

   (c) $\{x_n\} = \{\frac{1}{n}\}$

10. Look up the article mentioned after the proof of Theorem 7.1 and write a proof of the result stated there that every bounded sequence has a monotone subsequence.

11. Prove that if the sequence $\{x_n\}$ converges to x, then the arithmetic mean sequence also converges to x.

12. Give an example of a sequence $\{x_n\}$ which has 1 and 2 as limit points, and is such that the set $S = \{x_n : n \in \mathbb{N}\}$ has only 1 as a limit point.

13. In Example 8, find a formula for n_k, so that the subsequence $\{x_{n_k}\}$ converges to 1.

SECTION 8   CAUCHY SEQUENCES

We come now to the final theoretical concept in this chapter—that of a Cauchy sequence. It provides a method for deciding if a sequence converges, without knowing what the limit is. Whereas a sequence is called convergent if the terms get arbitrarily close to the limit value, a sequence will be called Cauchy if the terms eventually get arbitrarily close to each other. This is now stated more precisely.

Definition 1 $\{x_n\}$ is a *Cauchy sequence* if for every $\epsilon > 0$, there is some N so that $|x_n - x_m| < \epsilon$ for all $n, m > N$.

This sequence bears the name of Augustin Cauchy, a French mathematician who contributed to many areas of mathematics in the first half of the nineteenth century. One of his most influential discoveries was the idea of a rigorous definition for the limit concept as a basis for analysis. He introduced the widely used ϵ–δ (see Section 10) concept during his university lectures in 1821.

Example 1   We observe that $\{x_n\} = \{\frac{1}{n}\}$ is a Cauchy sequence. For every $\epsilon > 0$, there is some N so that $\frac{1}{N} < \epsilon$ by the Archimedean axiom. For $m > n > N$, $|\frac{1}{n} - \frac{1}{m}| < \frac{1}{n} \leq \frac{1}{N} < \epsilon$. Notice that no mention is made of the limit value of 0 for this sequence.   □

LEMMA 8.1   Every Cauchy sequence is bounded.

Proof.   Let $\{x_n\}$ be a Cauchy sequence. For $\epsilon = 1$, there is some N so that $|x_n - x_m| < 1$ for $n, m > N$. In particular, for $m = N$, we have
$$|x_n| - |x_N| \leq |x_n - x_N| < 1$$

Thus $1 + |x_N|$ is a bound for all the $|x_n|$ terms except possibly a finite number. Then a bound for all the $|x_n|$ terms is

$$B = \max\{|x_1|, |x_2|, \ldots, |x_{N-1}|, |x_N| + 1\} \quad \blacksquare$$

With the use of this lemma and the Bolzano-Weierstrass Theorem, we are now able to prove the following result.

THEOREM 8.2   A sequence $\{x_n\}$ is a Cauchy sequence if and only if it is a convergent sequence.

Proof of $\Leftarrow$:   This is the easier half of the proof. It follows directly from Definitions 4.2 and 8.1, and is much like the proofs given for Theorem 4.1. Let $\epsilon > 0$ be given. Since $\{x_n\}$ converges, we may suppose that $\lim x_n = x$. Then there is some N so that $|x_n - x| < \frac{\epsilon}{2}$ for all $n > N$. For $n, m > N$ we have

$$|x_n - x_m| = |x_n - x + x - x_m|$$

$$\leq |x_n - x| + |x - x_m|$$

$$< \frac{\epsilon}{2} + \frac{\epsilon}{2} = \epsilon \quad \blacksquare$$

Proof of $\Rightarrow$:   If $\{x_n\}$ is a Cauchy sequence, it must be bounded because of Lemma 8.1. Thus it must have a convergent subsequence by Theorem 7.1. If we let $\{x_{n_k}\}$ denote this convergent subsequence, and let x_o denote its limit, then we need only show that $\lim_{n \to \infty} x_n = x_o$.

Let $\epsilon > 0$ be given. There is some N_1, so that if $n, m > N_1$, then $|x_n - x_m| < \frac{\epsilon}{2}$. There is some N_2, so that if $n_k > N_2$, then $|x_{n_k} - x_o| < \frac{\epsilon}{2}$. Let $N = \max\{N_1, N_2\}$, let k be such that $n_k > N$, and let n be any natural number so that $n > N$. Then

$$|x_n - x_o| = |x_n - x_{n_k} + x_{n_k} - x_o|$$

$$\leq |x_n - x_{n_k}| + |x_{n_k} - x_o|$$

$$< \frac{\epsilon}{2} + \frac{\epsilon}{2} = \epsilon \quad \blacksquare$$

Example 2   Let $x_n = \frac{1}{2}(x_{n-1} + x_{n-2})$ where $x_1 = 1$ and $x_2 = 2$. We proved in Section 7 that this sequence converged by using Corollary 7.3. We now prove that this sequence is a Cauchy sequence, and therefore it must converge because of Theorem 8.2.

Because of the averaging nature of the definition for this sequence, we have $|x_2 - x_1| = 1$, $|x_3 - x_2| = \frac{1}{2}$, $|x_4 - x_3| = \frac{1}{4}$, and in general $|x_n - x_{n-1}| = \frac{1}{2^{n+2}}$.

If $m < n$, we have

$$|x_n - x_m| \;=\; |x_n - x_{n-1} + x_{n-1} + \cdots - x_{m+1} + x_{m+1} - x_m|$$

$$\leq \sum_{i=m+1}^{n} |x_i - x_{i-1}|$$

$$\leq \sum_{i=m+1}^{n} \frac{1}{2^{i+2}}$$

$$= \frac{1}{2^{m+3}}\left(1 + \frac{1}{2} + \cdots + \frac{1}{2^{n-m-1}}\right)$$

$$< \frac{1}{2^{m+3}} \cdot 2 \;=\; \frac{1}{2^{m+2}}$$

Since there is some M, so that $m > M$ implies $\frac{1}{2^{m+2}} < \epsilon$, it follows that this is a Cauchy sequence.   □

Comment. This entire chapter has been very theoretical in nature and may have been painful at times for you, because of the preciseness required while working with concepts and methods that were undoubtedly new, even to someone who has taken the entire calculus course. However, a serious effort to gain familiarity with these ideas will pay rich dividends in the understanding of the material in future chapters. It will be of benefit to take time for a brief review of the material covered so far in this chapter.

So far, there have been 11 definitions and 7 theorems in this chapter. This is not an unreasonable number for you to memorize, and this process of memorization should reinforce your understanding as well. Through such a review, you may notice that a number of the definitions have a similar form—they can be stated using the phrase "every interval about x contains."

Definition 4.2 x is *the limit* of the convergent sequence $\{x_n\}$ if "every interval about x contains" all but a finite number of the x_n terms.

Definition 5.1 x is the *least upper bound* of the non-empty set S if "every interval about x contains" at least one element from S that is less than or equal to x, and no elements from S that are greater than x.

Definition 5.5 x is a *cluster point* for the set S if "every interval about x contains" an element from S that is not equal to x.

Definition 7.2 x is a *limit point* for the sequence $\{x_n\}$ if "every interval about x contains" an infinite number of the x_n terms.

You are asked in the exercises to add the glb definition to this list. Each of these definitions can be stated in an equivalent form using the ϵ notation, which is the form in which they were originally stated earlier in the text.

In addition to this overview of the main definitions, we can also make the following flow chart involving the theorems of this chapter. The lines in the chart indicate the logical order in which the proofs are derived. The dotted boxes indicate the results that will be covered in later chapters and that are consequences of the theorems proved in Chapter 2.

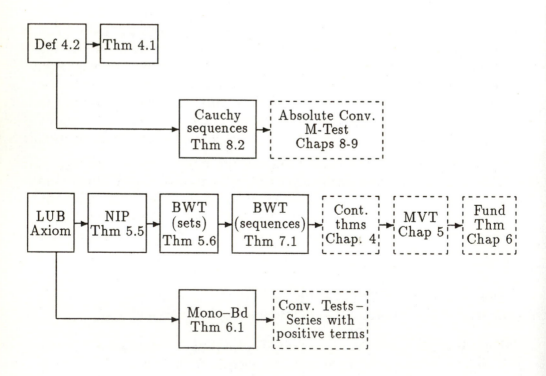

Figure 8.1

In case you feel you should not go on to Chapter 3 until the ideas from Chapter 2 are mastered, a general property of mathematical learning is now stated. It would be possible for you to spend several more days of study on the ideas in Chapter 2 and still have a shaky level of comprehension. However, by going on to the new concepts in Chapter 3, you will find that the ideas that seem so foreign now will soon appear much more elementary and understandable. Of course, the ideas in Chapter 3 will still be unclear, but that can be resolved by pressing on to Chapter 4.

EXERCISES 8

1. Prove directly from Definition 1 that the sum and product of two Cauchy sequences is again a Cauchy sequence. Pattern the form of your proofs after those of Theorem 4.1.

2. Verify that the following are Cauchy sequences, by using Definition 1.

    (a) $x_n = \frac{n+2}{n}$

    (b) $x_n = \frac{n}{n+2}$

    (c) $x_n = 1 + \frac{1}{4} + \frac{1}{9} + \cdots + \frac{1}{n^2}$

3. Apply Definition 1 to the sequence of partial sums $\{S_n\}$ for the series $\sum x_n$, in order to obtain an expression using the x_n terms that is equivalent to $|S_n - S_m| < \epsilon$. See Definition 9.1 if needed. This is called the Cauchy convergence criterion, which will be used in Chapters 8 and 9.

4. Write the greatest lower bound definition in the terminology used for the other definitions summarized at the end of this section.

SECTION 9   SEQUENCES OF PARTIAL SUMS

Because this special type of sequence is so important in the study of analysis, we devote a section to it here as a preview of the material in Chapter 8.

Definition 1   If $\{x_n\}$ is a sequence of real numbers, then $\{S_n\}$, where $S_n = x_1 + x_2 + \ldots + x_n$, is called the *sequence of partial sums* determined by $\{x_n\}$.

The concept of an infinite series can then be introduced by identifying the infinite series $\sum_{n=1}^{\infty} x_n$ with the sequence of partial sums $\{S_n\}$. Thus an infinite series is simply an alternate way to refer to the sequence $\{S_n\}$, and yet the concept of a series takes on a life of its own, becoming perhaps the largest and most useful part of analysis. $\sum x_n$ will mean $\sum_{n=1}^{\infty} x_n$ throughout.

Definition 2   The infinite series $\sum_{n=1}^{\infty} x_n$ *converges to* x, if the sequence of partial sums $\{S_n\}$, where $S_n = \sum_{i=1}^{n} x_i$ converges to x.

For example, the sequence $\{x_n\} = \{\frac{1}{n}\}$ determines the sequence of partial sums $\{S_n\} = \{1 + \frac{1}{2} + \ldots + \frac{1}{n}\}$, or the series $\sum_{n=1}^{\infty} \frac{1}{n}$, while the sequence $\{x_n\} = \{\frac{(-1)^{n+1}}{n^2}\}$ determines the sequence of partial sums $\{S_n\} = \{1 - \frac{1}{4} + \frac{1}{9} - \cdots + \frac{(-1)^{n+1}}{n^2}\}$, or the series $\sum_{n=1}^{\infty} \frac{(-1)^{n+1}}{n^2}$.

LEMMA 9.1   If the sequence $\{S_n\}$ of partial sums for the series $\sum_{n=1}^{\infty} x_n$ converges, then the sequence $\{x_n\}$ converges to 0.

Proof.   Suppose that $\lim_{n \to \infty} S_n = S$.

Since $x_n = S_n - S_{n-1}$, it follows that

$$\lim_{n \to \infty} x_n = \lim_{n \to \infty} (S_n - S_{n-1}) = \lim_{n \to \infty} S_n - \lim_{n \to \infty} S_{n-1} = S - S = 0 \quad \blacksquare$$

An equivalent result to Lemma 9.1 can be stated using series symbolism as follows.

COROLLARY 9.2 (nth Term Test)

If the series $\sum_{n=1}^{\infty} x_n$ converges, then $\lim_{n \to \infty} x_n = 0$.

The contrapositive of this can then be used as a test for divergence.

COROLLARY 9.3

If $\lim_{n \to \infty} x_n \neq 0$, then $\sum_{n=1}^{\infty} x_n$ diverges.

We next consider some examples of well-known infinite series.

Example 1 $\sum_{n=0}^{\infty} r^n = 1 + r + r^2 + \cdots$, where r is a constant, is called a *geometric series*, with ratio r. This series converges to $\frac{1}{1-r}$ if $0 < |r| < 1$, and diverges if $|r| \geq 1$. $\square$

THEOREM 9.4 The geometric series $\sum_{n=0}^{\infty} r^n$ converges to the sum $\frac{1}{1-r}$ if $0 < |r| < 1$, and diverges if $r \geq 1$.

Proof. We begin with the expression for the sequence of partial sums

$$S_n = 1 + r + r^2 + \cdots + r^n$$

Then
$$r\, S_n = r + r^2 + \cdots + r^n + r^{n+1}$$

and
$$(1 - r)\, S_n = 1 - r^{n+1}$$

so that
$$\lim_{n \to \infty} S_n = \lim_{n \to \infty} \frac{1 - r^{n+1}}{1 - r} = \frac{1}{1 - r} \quad \text{if } |r| < 1$$

by referring to Exercise 4.15, where part of the problem was to show that $\lim_{n \to \infty} c^n = 0$, for $|c| < 1$.

The case where $|r| \geq 1$ is left as an exercise. $\square$

LEMMA 9.5 If each $x_n > 0$, then the series $\sum x_n$ converges if and only if the sequence of partial sums $\{S_n\}$ is bounded above.

Proof. This is left as an exercise.

Comment. The *p-series* is defined as $\sum_{n=1}^{\infty} \frac{1}{n^p}$, where p is a positive constant. The *p*-series converges if $p > 1$, and diverges if $p \leq 1$. Finding values for a convergent *p*-series is a well-known problem in analysis, which is still not completely solved. When $p = 1$, the series $\sum_{n=1}^{\infty} \frac{1}{n}$ is called the *harmonic series*, and can be shown to diverge to $+\infty$. When $p = 2$, the series $\sum_{n=1}^{\infty} \frac{1}{n^2}$ can be shown to converge (see Example 2), but it is considerably more difficult to show that it converges to $\frac{\pi^2}{6}$.

Example 2   Prove that the series $\sum_{n=1}^{\infty} \frac{1}{n^2}$ converges.

Solution.   Since $x_n > 0$ for all n, the sequence of partial sums $\{S_n\}$, where $S_n = 1 + \frac{1}{2^2} + \cdots + \frac{1}{n^2}$ (and therefore the series $\sum \frac{1}{n^2}$) will converge if and only if $\{S_n\}$ is bounded above, by using Lemma 9.5. So we can show that $\sum \frac{1}{n^2}$ converges by finding an upper bound for the S_n terms. Here are two ways to do this.

Method 1 (Using an improper integral)

We compare the area of rectangles determined by $\sum_{n=1}^{\infty} \frac{1}{n^2}$ with the area under the curve $f(x) = \frac{1}{x^2}$. See Example 8 and Figure 2.2 in Section 2 for more details. Then

$$S_n = 1 + \frac{1}{2^2} + \cdots + \frac{1}{n^2}$$

$$< 1 + \int_1^n \frac{1}{x^2}\, dx$$

$$= 1 + \left. \frac{-1}{x} \right|_1^n = 2 - \frac{1}{n} < 2$$

Since S_n is bounded above by 2, $\sum \frac{1}{n^2}$ converges, and to a value that is less than or equal to 2. $\square$

Method 2 (Using a grouping approach)

$$S_{2^k - 1} = 1 + \left(\frac{1}{2^2} + \frac{1}{3^2} \right) + \left(\frac{1}{4^2} + \cdots + \frac{1}{7^2} \right) + \cdots$$

$$\cdots + \left(\frac{1}{(2^{k-1})^2} + \cdots + \frac{1}{(2^k - 1)^2} \right)$$

$$< \ 1 + 2 \cdot \frac{1}{2^2} + 4 \cdot \frac{1}{4^2} + \cdots + 2^{k-1} \cdot \frac{1}{(2^{k-1})^2}$$

$$= \ 1 + \frac{1}{2} + \frac{1}{4} + \cdots + \frac{1}{2^{k-1}}$$

$$= \ \frac{1 - \frac{1}{2^k}}{1 - \frac{1}{2}} \ = \ 2\left(1 - \frac{1}{2^k}\right) \ < \ 2$$

For all n, there is a value for k so that $n < 2^k - 1$, and we therefore have $S_n < S_{2^k-1} < 2$. Thus S_n converges, and to a value $S \leq 2$. $\quad\square$

On occasion, we depart from the strict use of the sequence of partial sums concept for establishing convergence of a series. We do this to simplify the notation, when it does not affect the outcome. This informal technique was used very effectively by the great analyst Euler during the eighteenth century, and we illustrate it below, using the same problem as in Method 2 above.

Method 2 (An alternate approach)

$$\sum_{n=1}^{\infty} \frac{1}{n^2} \ = \ 1 + \left(\frac{1}{2^2} + \frac{1}{3^2}\right) + \left(\frac{1}{4^2} + \cdots + \frac{1}{7^2}\right) + \left(\frac{1}{8^2} + \cdots + \frac{1}{15^2}\right) + \cdots$$

$$< \ 1 + 2 \cdot \frac{1}{2^2} + 4 \cdot \frac{1}{4^2} + 8 \cdot \frac{1}{8^2} + \cdots$$

$$= \ 1 + \frac{1}{2} + \frac{1}{4} + \frac{1}{8} + \cdots \ = \ \frac{1}{1 - \frac{1}{2}} \ = \ 2 \quad\square$$

THEOREM 9.6 The p-series $\sum_{n=1}^{\infty} \frac{1}{n^p}$ converges if $p > 1$, and diverges if $p \leq 1$.

Proof. We provide an outline of the proof for the case where $p > 1$, and leave it as an exercise for you to fill in the missing details. The proof resembles very closely the steps in Method 2 of Example 2.

$$\sum_{n=1}^{\infty} \frac{1}{n^p} = 1 + \left(\frac{1}{2^p} + \frac{1}{3^p}\right) + \left(\frac{1}{4^p} + \cdots + \frac{1}{7^p}\right)$$

$$+ \left(\frac{1}{8^p} + \cdots + \frac{1}{15^p}\right) + \cdots + \left(\frac{1}{(2^{k-1})^p} + \cdots + \frac{1}{(2^k - 1)^p}\right)$$

$$\leq 1 + 2 \cdot \frac{1}{2^p} + 4 \cdot \frac{1}{4^p} + 8 \cdot \frac{1}{8^p} + \cdots + 2^{k-1} \cdot \frac{1}{(2^{k-1})^p}$$

$$= 1 + \frac{1}{2^{p-1}} + \frac{1}{(2^{p-1})^2} + \frac{1}{(2^{p-1})^3} + \cdots + \frac{1}{(2^{p-1})^{k-1}}$$

$$= \frac{1 - \frac{1}{(2^{p-1})^k}}{1 - \frac{1}{2^{p-1}}} < \frac{2^{p-1}}{2^{p-1} - 1} \qquad \blacksquare$$

If $p \leq 0$, then $\lim_{n \to \infty} \frac{1}{n^p} \neq 0$, and so $\sum_{n=1}^{\infty} \frac{1}{n^p}$ diverges by the nth term test.

If $0 < p \leq 1$, we first observe that $\sum_{n=0}^{\infty} \frac{1}{(2^{p-1})^n}$ is a divergent geometric series, according to Theorem 9.4. Then

$$\sum_{n=1}^{\infty} \frac{1}{n^p} = 1 + \frac{1}{2^p} + \left(\frac{1}{3^p} + \frac{1}{4^p}\right) + \left(\frac{1}{5^p} + \cdots + \frac{1}{8^p}\right) + \cdots$$

$$> 1 + \frac{1}{2^p} + 2 \cdot \frac{1}{4^p} + 4 \cdot \frac{1}{8^p} + \cdots$$

$$> 1 + \frac{1}{2^p} + \frac{1}{2^{2p-1}} + \frac{1}{2^{3p-2}} + \cdots$$

$$= 1 + \frac{1}{2^p}\left(1 + \frac{1}{2^{p-1}} + \frac{1}{(2^{p-1})^2} + \frac{1}{(2^{p-1})^3} + \cdots\right)$$

Since $\sum_{n=1}^{\infty} \frac{1}{n^p}$ is greater than an expression containing a series that diverges to $+\infty$, the series $\sum_{n=1}^{\infty} \frac{1}{n^p}$ will diverge to $+\infty$ as well. See also Exercise 8. $\blacksquare$

Example 3 $\sum_{n=1}^{\infty} \frac{(-1)^{n+1}}{n} = 1 - \frac{1}{2} + \frac{1}{3} - \frac{1}{4} + - \cdots$ is called the *alternating harmonic series*. We show in Chapter 8 that it converges by the Alternating Series Test, and in Chapter 9 that it converges to $\ln 2$ by using a Maclaurin series expansion. $\square$

If desired, the material in Chapter 8 could be studied next. Chapter 8 contains the well-known convergence tests for series of constants. The proofs for all these tests follow from the results on sequences obtained in this chapter. We illustrate this by considering a proof of the comparison test.

THEOREM 9.7 (Comparison Test)

If $\sum y_n$ converges and $0 \leq x_n \leq y_n$ for all n, then $\sum x_n$ converges. In addition, if $\sum y_n$ converges to y and $\sum x_n$ converges to x, then $x \leq y$.

Proof. Let $Y_n = y_1 + y_2 + \cdots + y_n$ and $X_n = x_1 + x_2 + \cdots + x_n$. Since $X_n \leq Y_n$, and the sequence $\{Y_n\}$ is bounded, the sequence $\{X_n\}$ is also bounded, and therefore convergent. It also follows that $x = \lim X_n \leq \lim Y_n = y$. ∎

Example 4 The series

$$\sum_{n=1}^{\infty} y_n = \sum_{n=1}^{\infty} \frac{1}{2^{n-1}} = 1 + \frac{1}{2} + \frac{1}{4} + \cdots$$

is a convergent geometric series. The series

$$\sum_{n=1}^{\infty} x_n = \sum_{n=1}^{\infty} \frac{1}{1 + 2^n} = \frac{1}{3} + \frac{1}{5} + \frac{1}{9} + \cdots$$

converges by the Comparison Test since $\frac{1}{1+2^n} < \frac{1}{2^{n-1}}$ for all n. The series $\sum y_n$ cannot be used to prove that $\sum_{n=1}^{\infty} \frac{1}{n^3}$ converges because although $\frac{1}{n^3} \leq \frac{1}{2^{n-1}}$ for $n \leq 11$, the inequality is reversed for $n > 11$. Note that this does not mean that $\sum \frac{1}{n^3}$ diverges. □

EXERCISES 9

1. Find the value for each of the following geometric series.
   (a) $1 + \frac{1}{3} + \frac{1}{9} + \frac{1}{27} + \cdots$
   (b) $1 - \frac{1}{4} + \frac{1}{16} - \frac{1}{64} + - \cdots$
   (c) $3 + \frac{6}{5} + \frac{12}{25} + \cdots$
2. Prove that $\sum_{n=1}^{\infty} \frac{1}{n^3}$ converges, using the two methods given in Example 2.

3. Adjust the two methods in Example 2 in order to prove that $\sum_{n=1}^{\infty} \frac{1}{n}$ diverges to $+\infty$.

4. Use an improper integral approach to prove that the geometric series $\sum_{n=0}^{\infty} r^n$ diverges if $|r| \geq 1$.

5. Prove Lemma 9.5.

6. Provide the missing steps and reasons in the proof of the convergence of the p-series, when $p > 1$, given in the proof of Theorem 9.6.

7. Write out the first 10 terms for the sequence of partial sums $\{S_n\}$ for the series $\sum_{n=1}^{\infty} \frac{(-1)^{n+1}}{n}$. What do you observe about this sequence that might be useful in proving its convergence?

8. State and prove a comparison test analog to Theorem 9.7 that can be used to prove that a given series diverges.

9. Use the Comparison Test and the p-series results in Theorem 9.6 to determine whether the following series converge or diverge. You may conclude that some of these cannot be decided yet.

   (a) $\sum_{n=0}^{\infty} \frac{1}{2^n+2} = \frac{1}{3} + \frac{1}{4} + \frac{1}{6} + \frac{1}{10} + \cdots$

   (b) $\sum_{n=1}^{\infty} \frac{1}{n^2+1} = \frac{1}{2} + \frac{1}{5} + \frac{1}{10} + \frac{1}{17} + \cdots$

   (c) $\sum_{n=2}^{\infty} \frac{1}{n^2-1} = \frac{1}{3} + \frac{1}{8} + \frac{1}{15} + \frac{1}{24} + \cdots$

   (d) $\sum_{n=0}^{\infty} \frac{1}{n!} = 1 + \frac{1}{2} + \frac{1}{6} + \frac{1}{24} + \cdots$

   (e) $\sum_{n=1}^{\infty} \frac{1}{2n} = \frac{1}{2} + \frac{1}{4} + \frac{1}{6} + \frac{1}{8} + \cdots$

   (f) $\sum_{n=1}^{\infty} \frac{1}{n+2} = \frac{1}{3} + \frac{1}{4} + \frac{1}{5} + \frac{1}{6} + \cdots$

3

Limits of Functions

SECTION 10   DEFINITION FOR THE LIMIT OF A FUNCTION

Throughout this book, when we speak of a function, we will mean a mapping f from a domain set A into a codomain set B, and we represent this by $f : A \to B$. To say that f is a real-valued function means that $B \subset \mathbf{R}$. Unless otherwise stated, for functions studied in this book, A and B are subsets of the set of real numbers.

In Chapter 2, we made a fairly complete study of the limit concept as seen in the setting of sequences where $\lim_{n \to \infty} x_n = x$. Among other things, we learned how to work with the ϵ–N notation. In the next several chapters, we will work with the ϵ–δ symbolism as it is expressed in the setting of $\lim_{x \to c} f(x) = L$. We devote Chapter 3 to the task of gaining familiarity with some rather mechanical skills, so that we will be free to concentrate on the properties of continuity in Chapter 4. This section contains the limit definition and Limit Theorem for functions, which closely resemble the limit definition and Limit Theorem for sequences that was presented in Section 4.

Definition 1   Let $f : A \to \mathbf{R}$, let c be a cluster point of A, and let L be a real number such that for every $\epsilon > 0$, there is some $\delta > 0$, so that for every $x \neq c$ in the domain of f for which $|x - c| < \delta$, it follows that $|f(x) - L| < \epsilon$. In this case, we say that the function f has the limit value L as x approaches c, and express this by the symbol $\lim_{x \to c} f(x) = L$.

We can use the symbol D_f to represent the set of domain values for the function f, and the symbol R_f to represent the set of range values; that is, $R_f = \{f(x) : x \in D_f\}$. We sometimes use the symbol δ_ϵ or $\delta(\epsilon)$ to emphasize the fact that the value for δ depends upon the value of ϵ.

Example 1   Apply Definition 1 to $\lim_{x \to 2}(2x^2 + 3x - 6)$.

Solution.   Since x approaches 2, we start our work with the conjecture that

$$L = f(2) = 2 \cdot 2^2 + 3 \cdot 2 - 6 = 8$$

We may assume that $\delta \leq 1$. Then since $|x - 2| < \delta$, we know that $1 < x < 3$ and $9 < 2x + 7 < 13$. Then

$$|f(x) - L| = |(2x^2 + 3x - 6) - 8| = |2x + 7||x - 2| < 13\delta$$

by using the above comments. To make $|f(x) - L|$ less than ϵ, we can define $\delta_\epsilon = \min\{1, \frac{\epsilon}{13}\}$. While this is the usual expression for δ_ϵ, there are many other possibilities. Two of these are $\delta_\epsilon = \min\{2, \frac{\epsilon}{15}\}$ and $\delta_\epsilon = \min\{\frac{1}{2}, \frac{\epsilon}{12}\}$.   □

Example 2   Apply Definition 1 to $\lim_{x \to -5}(2x^2 + 3x - 6)$.

Solution.   We first conjecture that $L = f(-5) = 29$. Then

$$|(2x^2 + 3x - 6) - 29| = |2x^2 + 3x - 35| = |2x - 7||x + 5| < 19\delta$$

since for $\delta \leq 1$, $-6 < x < -4$ and $-19 < 2x - 7 < -15$. So let $\delta_\epsilon = \min\{1, \frac{\epsilon}{19}\}$.   □

Example 3   Apply Definition 1 to $\lim_{x \to 1} \frac{1}{x^2}$.

Solution.   In this case, it will not work to only assume that $\delta \leq 1$ or $0 < x < 2$, because this will not provide a lower bound for the x term that occurs in the denominator. However, we can assume that $\delta \leq \frac{1}{2}$ holds, which yields $\frac{1}{2} < x < \frac{3}{2}$ so that $\frac{1}{x} < 2$. Now

$$|f(x) - L| = \left|\frac{1}{x^2} - 1\right| = \left|\frac{(1 - x)(1 + x)}{x^2}\right| < 2^2 \cdot \frac{5}{2} \cdot \delta < \epsilon$$

if we define $\delta_\epsilon = \min\{\frac{1}{2}, \frac{\epsilon}{10}\}$.   □

We next prove a lemma about boundedness that serves as an analog to Lemmas 4.2 and 4.3. The form of the statement will be somewhat different because there is no function analog for the phrase "for all but a finite number of n" that occurred in our discussion of sequences.

If $\lim_{x \to c} f(x) = L$, we can easily show that $\lim_{x \to c} |f(x)| = |L|$ (see the exercises). Then for $\epsilon = 1$, there is some δ_1 so that

$$|L| - 1 \; < \; |f(x)| \; < \; |L| + 1$$

for all x so that $|x - c| < \delta_1$. For $\epsilon = \frac{|L|}{2}$, there is some δ_2 so that

$$\frac{|L|}{2} \; < \; |f(x)| \; < \; \frac{3|L|}{2}$$

for all x so that $|x - c| < \delta_2$. We normally use the first bound with its tolerance of 1 unless the value of L is such that $-1 < L < 1$, and we need a bound for $\frac{1}{|f(x)|}$ in our work. In this case, we choose the second tolerance of $\frac{|L|}{2}$, so that $\frac{|L|}{2} < |f(x)|$, or $\frac{1}{|f(x)|} < \frac{2}{|L|}$. These comments furnish the details of a proof for the following lemma.

LEMMA 10.1   If $\lim_{x \to c} f(x) = L$, there is some δ so that if $|x - c| < \delta$, then $\frac{|L|}{2} \; < \; |f(x)| \; < \; |L| + 1$.

Definition 1 and the expression of boundedness in Lemma 10.1 now enable us to prove the following Limit Theorem for functions.

THEOREM 10.2   Let f and g be defined on the set A, and let c be a cluster point of A. If $\lim_{x \to c} f(x) = L$ and $\lim_{x \to c} g(x) = M$, then the following results are true.

(1) $\lim_{x \to c} f(x) + g(x) \; = \; L + M$.

(2) $\lim_{x \to c} f(x)g(x) \; = \; LM$.

(3) $\lim_{x \to c} \frac{f(x)}{g(x)} \; = \; \frac{L}{M}$ if $M \neq 0$ and $g(x) \neq 0$ for all x in some neighborhood of c.

The proofs of these results closely resemble the form of the proofs of the corresponding results for convergent sequences from Section 4. Only the proof of one case will be shown here, with the others left as exercises.

Proof of (3).   Let $\epsilon > 0$ be given.
There is some $\delta_1 > 0$, so that for $|x - c| < \delta_1$, $\frac{|M|}{2} < |g(x)|$ or $\frac{1}{|g(x)|} < \frac{2}{|M|}$.
There is some $\delta_2 > 0$, so that for $|x - c| < \delta_2$, $|f(x) - L| < \frac{\epsilon |M|}{4}$.
There is some $\delta_3 > 0$, so that for $|x - c| < \delta_3$, $|g(x) - M| < \frac{\epsilon |M|^2}{4|L|+1}$.

Let $\delta_\epsilon = \min\{\delta_1, \delta_2, \delta_3\}$, and choose x so that $|x - c| < \delta_\epsilon$.

$$\text{Then} \quad \left|\frac{f(x)}{g(x)} - \frac{L}{M}\right| = \left|\frac{M\,f(x) - L\,g(x)}{M\,g(x)}\right|$$

$$= \left|\frac{M\,f(x) - LM + LM - L\,g(x)}{Mg(x)}\right|$$

$$\leq \frac{|f(x) - L|}{|g(x)|} + \frac{|L|\,|g(x) - M|}{|M|\,|g(x)|}$$

$$< \frac{\epsilon|M|}{4} \cdot \frac{2}{|M|} + \frac{2|L|}{|M|^2} \cdot \frac{\epsilon|M|^2}{4|L| + 1}$$

$$< \frac{\epsilon}{2} + \frac{\epsilon}{2} = \epsilon \quad \blacksquare$$

You should find it instructive to work backwards to show why the particular bounds of $\frac{2}{|M|}$, $\frac{\epsilon|M|}{4}$, and $\frac{\epsilon|M|^2}{4|L|+1}$ were chosen in this proof. See the proof of Theorem 4.1 for a hint if needed. Note also that since $|x - c| < \delta_\epsilon$, then $|x - c| < \delta_1$, $|x - c| < \delta_2$, and $|x - c| < \delta_3$, so that all the above bounds apply.

We next present the definitions for a one-sided limit and a monotone function, and then prove a lemma using these concepts, which will be useful later in the book.

Definition 2   The *right-hand limit* is denoted by $\lim_{x \to c+} f(x) = L$ and means that for every $\epsilon > 0$, there is some $\delta > 0$ so that for every $x \in D_f$ for which $c < x < c + \delta$, it follows that $|f(x) - L| < \epsilon$. Similarly, the *left-hand limit* is denoted by $\lim_{x \to c-} f(x) = L$, and means that for every $\epsilon > 0$, there is some $\delta > 0$ so that for every $x \in D_f$ for which $c - \delta < x < c$, it follows that $|f(x) - L| < \epsilon$.

By comparing Definition 2 with Definition 1, we are able to state that $\lim_{x \to c} f(x)$ exists and equals L if and only if both one-sided limits exist and equal L.

Definition 3   The function f is *increasing on the interval I* if we have $f(x) < f(y)$ for all $x < y$ in I. f is *decreasing on I* if $f(x) > f(y)$ whenever $x < y$ in I. f is *nondecreasing on I* if $f(x) \le f(y)$ for all $x < y$ in I and *nonincreasing on I* if $f(x) \ge f(y)$ for all $x < y$ in I. A function satisfying any of the above conditions is called *monotone on I*. Notice that an increasing function is nondecreasing, but the converse does not hold.

Example 4   The function $\sin x$ is increasing on the interval $[0, \frac{\pi}{2}]$ and decreasing on the interval $[\frac{\pi}{2}, \frac{3\pi}{2}]$. The greatest integer function $[x]$ is nondecreasing, but not increasing, on the interval $[0, 3]$. $\square$

Definition 4   Let the function $f : A \to \mathbf{R}$ be defined. f is *bounded above on A* if there is a real number $b > 0$, so that $f(x) \le b$ for all $x \in A$ and *bounded below on A* if $b \le f(x)$ for all $x \in A$. f is *bounded on A* if $|f(x)| \le b$ for all $x \in A$. A bounded function is bounded above and below, since $|f(x)| \le b$ is equivalent to $-b \le f(x) \le b$.

Example 5   The function $f(x) = x^3$ is bounded above on the interval $(-\infty, -2)$ since $x^3 \le -8$ for $x \le -2$, and bounded below on $[0, \infty)$, since $0 \le x^3$ for $x \ge 0$. $f(x) = x^3$ is bounded on sets of the form $[-a, a]$, but not on $(-\infty, \infty)$. $\square$

LEMMA 10.3   If f is nondecreasing and bounded above on the interval (a, b), then $\lim_{x \to b-} f(x)$ exists in $\mathbf{R}$. If f is nonincreasing and bounded below on the interval (a, b), then $\lim_{x \to b-} f(x)$ exists in $\mathbf{R}$.

Proof.   Let $S = \{ f(x) : x \in (a, b) \}$. Since S is bounded above, the real number $\sup S$ exists by the LUB axiom, and we denote it by $t = \sup S$. We assert that $\lim_{x \to b-} f(x) = t$. For if $\epsilon > 0$, there is some $x_\epsilon \in (a, b)$, so that $t - \epsilon < f(x_\epsilon) \le t$ by Lemma 5.2. Also, $f(x_\epsilon) \le f(x) \le t$ for all $x_\epsilon < x < b$. By letting $\delta_\epsilon = b - x_\epsilon$, we have proved that $\lim_{x \to b-} f(x) = t$ using Definition 2.

The proof of the second half of this lemma is left as an exercise.   ∎

THEOREM 10.4 (Squeeze Theorem)

Let I be an interval containing the point c, and assume $f(x) \le g(x) \le h(x)$ holds for all $x \in I$. If $\lim_{x \to c} f(x) = L = \lim_{x \to c} h(x)$, then $\lim_{x \to c} g(x) = L$ also.

Proof. This is left as an exercise.

We notice that the Limit Theorem is much broader than as stated in Theorem 10.2. For example, we would assert that $\lim_{x \to 0} \frac{1}{x^2} = +\infty$, even though this case is not covered by Theorem 10.2, since we have a denominator where $M = 0$. We can think of $\frac{1}{0}$ as a *form*, which is always infinite, although there is some uncertainty between $+\infty$ and $-\infty$, in a case such as $\lim_{x \to 0} \frac{1}{x}$. In a similar way, the form $\frac{1}{\infty}$ can always be taken as zero, the form $\infty + \infty$ will equal ∞, and $k \cdot \infty$ will have the value of $-\infty$ for $k < 0$. On the other hand, forms as $\frac{0}{0}$ and $\infty - \infty$ will be considered as undetermined, because they do not have the same value for all functions f and g. In calculus, these are called *indeterminate forms*. In order to state these ideas more precisely, we need to extend the definition of limit to include the possibility that c and/or L could be infinite. This is the purpose of the next section.

EXERCISES 10

1. Illustrate Definition 1 for the following limits by finding an appropriate expression for δ_ϵ.

   (a) $\lim_{x \to 4} (x^2 - 3x + 2)$

   (b) $\lim_{x \to -2} (x^3 + 3x^2 - 2x + 4)$

   (c) $\lim_{x \to t} (ax^2 + bx + c)$

   (d) $\lim_{x \to 1} \frac{1}{x^2 + 4}$

   (e) $\lim_{x \to 3} \frac{1}{4 - x^2}$

   (f) $\lim_{x \to 2} \sqrt{x + 2}$

   (g) $\lim_{x \to c} \frac{1}{x^2}$

2. Prove the sum result (1) of Theorem 10.2.

3. Prove the product result (2) of Theorem 10.2.

4. Answer the question about the 3 bounds that was stated in the comment after the proof of result (3) of Theorem 10.2.

5. Use Definition 1 to prove that $\lim_{x \to c} |f(x)| = |L|$ if $\lim_{x \to c} f(x) = L$. Give an example to show that the converse is not always true.

6. Prove Theorem 10.4.

7. Prove the second statement in Lemma 10.3.

8. Prove that if f is nondecreasing and bounded above on the interval $I = (a, \infty)$, then $\lim_{x \to \infty} f(x)$ exists in $\mathbf{R}$. Compare this result with Lemma 10.3.

SECTION 11   THE GENERAL LIMIT THEOREM

Definition 10.1 for $\lim_{x \to c} f(x) = L$ assumed that both c and L were finite real numbers. To handle the cases where c and/or L approach $+\infty$ or $-\infty$ by increasing or decreasing beyond all finite bounds, we must consider the 9 separate cases in Definition 1 below. Some of the cases are done as a guide and others are left blank for you to complete. To standardize the notation, let ϵ represent a small range tolerance when L is finite, and let R represent a large positive range tolerance when L is infinite. Similarly, let δ represent a small domain tolerance when c is finite, and let D represent a large domain tolerance when c is infinite.

Definition 1

   (1) $\lim_{x \to c} f(x) = L$ means   For every $\epsilon > 0$, there is $\delta_\epsilon > 0$ so that $|f(x) - L| < \epsilon$ for $|x - c| < \delta_\epsilon$

   (2) $\lim_{x \to c} f(x) = +\infty$ means   For every $R > 0$, there is $\delta_R > 0$ so that $f(x) > R$ for $|x - c| < \delta_R$

   (3) $\lim_{x \to c} f(x) = -\infty$ means

   (4) $\lim_{x \to \infty} f(x) = L$ means   For every $\epsilon > 0$, there is $D_\epsilon > 0$ so that $|f(x) - L| < \epsilon$ for $x > D_\epsilon$

   (5) $\lim_{x \to \infty} f(x) = +\infty$ means

   (6) $\lim_{x \to \infty} f(x) = -\infty$ means   For every $R > 0$, there is $D_R > 0$ so that $f(x) < -R$ for $x > D_R$

   (7) $\lim_{x \to -\infty} f(x) = L$ means

   (8) $\lim_{x \to -\infty} f(x) = +\infty$ means   For every $R > 0$, there is $D_R > 0$ so that $f(x) > R$ for $x < -D_R$

   (9) $\lim_{x \to -\infty} f(x) = -\infty$ means   For every $R > 0$, there is $D_R > 0$ so that $f(x) < -R$ for $x < -D_R$

The general form for each part of the limit definition is that for every range tolerance, there is a corresponding domain tolerance, so that if the independent variable satisfies the domain tolerance, the value for the corresponding dependent variable satisfies the given range tolerance.

You are reminded that this chapter is devoted to the task of gaining technical competence with all forms of the limit definition, and as such may seem tedious at times. It may be viewed as comparable to the exercises that a pianist uses to loosen up before tackling a composition by Chopin or Rachmaninoff.

Example 1   Prove that $\lim_{x \to 0} \frac{1}{x^2} = \infty$.

Solution.   We use form (2) of Definition 1, which requires that we find δ_R so that $\frac{1}{x^2} > R$ when $|x - 0| < \delta_R$. But $\frac{1}{x^2} > R$ is equivalent to $x^2 < \frac{1}{R}$ or $x < \frac{1}{\sqrt{R}}$. Therefore we set $\delta_R = \frac{1}{\sqrt{R}}$.   □

Example 2   Prove that $\lim_{x \to \infty} \frac{1}{x^2} = 0$.

Solution.   We use form (4) of Definition 1, which requires that we find D_ϵ so that $|\frac{1}{x^2} - 0| < \epsilon$ for $x > D_\epsilon$. But $|\frac{1}{x^2} - 0| = \frac{1}{x^2} < \epsilon$ whenever $x^2 > \frac{1}{\epsilon}$ or when $x > \frac{1}{\sqrt{\epsilon}}$. So define $D_\epsilon = \frac{1}{\sqrt{\epsilon}}$.   □

Example 3   Prove that $\lim_{x \to -\infty} x^2 = \infty$.

Solution.   We will have $x^2 > R$ if $|x| > \sqrt{R}$, or in this case, if $x < -\sqrt{R}$. So using form (8) of the Definition 1, we can let $D_R = \sqrt{R}$.   □

Example 4   Prove that $\lim_{x \to 1^+} \frac{1}{x^2 - 1} = +\infty$.

Solution.   Given $R > 0$, since $x \to 1^+$, we know that $1 < x < 2$, so that $2 < x + 1 < 3$ and $\frac{1}{x+1} > \frac{1}{3}$. If we set $\delta_R = \min\{1, \frac{1}{3R}\}$, then $x - 1 < \delta_R$ implies that $x - 1 < \frac{1}{3R}$ or $\frac{1}{x-1} > 3R$. Finally

$$\frac{1}{x^2 - 1} = \frac{1}{x+1} \cdot \frac{1}{x-1} > \frac{1}{3}(3R) = R$$

and the limit definition is satisfied.   □

We now generalize the result in Example 2 to say that $\frac{k}{\infty}$ is a *well defined form* with the value of 0 because of the following result, which is true regardless of which functions are used for f and g.

THEOREM 11.1   If $\lim_{x \to c} f(x) = k > 0$ and $\lim_{x \to c} = g(x) = +\infty$, then $\lim_{x \to c} \frac{f(x)}{g(x)} = 0$.

Proof.   Let $\epsilon > 0$ be given. Since $\lim_{x \to c} f(x) = k > 0$, there is some δ_1 so that for $|x - c| < \delta_1$, we have $f(x) < k + 1$. Since $\lim_{x \to c} g(x) = +\infty$, there is some δ_2, so that if $|x - c| < \delta_2$, then $g(x) > \frac{k+1}{\epsilon}$ or $\frac{1}{g(x)} < \frac{\epsilon}{k+1}$. Let $\delta_\epsilon = \min\{\delta_1, \delta_2\}$. Choose x so that $|x - c| < \delta_\epsilon$. Then

$$\left| \frac{f(x)}{g(x)} - 0 \right| = f(x) \cdot \frac{1}{g(x)} < k + 1 \cdot \frac{\epsilon}{k+1} = \epsilon \quad \blacksquare$$

THEOREM 11.2   If $\lim_{x \to \infty} f(x) = k > 0$, $\lim_{x \to \infty} g(x) = 0$, and there is some $D_1 > 0$ so that $g(x) > 0$ for all $x > D_1$, then we have $\lim_{x \to \infty} \frac{f(x)}{g(x)} = +\infty$.

Proof.   We will use form (5) of Definition 1. Let $R > 0$ be given. There is some $D_2 > 0$, so that $f(x) > \frac{k}{2}$ for all $x > D_2$. There is some $D_3 > 0$, so that $g(x) < \frac{k}{2R}$ for all $x > D_3$. Choose $D_R = \max\{D_1, D_2, D_3\}$, and let $x > D_R$. Then

$$\frac{f(x)}{g(x)} = f(x) \cdot \frac{1}{g(x)} > \frac{k}{2} \cdot \frac{2R}{k} = R \quad \blacksquare$$

We next prove that the form $k \cdot \infty = -\infty$ if $k < 0$.

THEOREM 11.3   Let $\lim f(x) = k < 0$ and $\lim g(x) = +\infty$. Then $\lim f(x)g(x) = -\infty$.

Proof.   Let $R > 0$ be given. We want to show that $f(x)g(x) < -R$. Since $\lim f(x) = k < 0$, we can choose x so that $\frac{3k}{2} < f(x) < \frac{k}{2}$, or $0 < -\frac{k}{2} < -f(x) < -\frac{3k}{2}$. Since $\lim g(x) = +\infty$, we can make $g(x)$ greater than any positive bound, for appropriate values of x. We determine that $g(x) > \frac{-2R}{k}$ is the best choice. Then

$$-f(x)g(x) > \left(\frac{-k}{2} \right) \left(\frac{-2R}{k} \right) = R$$

or $f(x)g(x) < -R$, as was desired. $\blacksquare$

Comment.   You may at first feel that an error has been made in the statement and proof of Theorem 11.3, because no mention was included as to whether the limit is taken as x approaches c, $+\infty$, $-\infty$, or whether the limit is one-sided. However, you should be convinced that the essential part of the proof is not affected by such a choice.

Similar proofs are possible for the forms $\frac{1}{0} = \pm\infty$, $\infty + \infty = \infty$, $k\cdot\infty = \infty$ for $k > 0$ and $k\cdot\infty = -\infty$ for $k < 0$, $\infty(-\infty) = -\infty$, and other well-defined forms. On the other hand, some forms are indeterminate, because the value of the form depends upon the particular functions f and g. For example, the form $\frac{0}{0}$ is indeterminate because if $\lim_{x\to c} f(x) = 0$ and $\lim_{x\to c} g(x) = 0$, then $\lim_{x\to c} \frac{f(x)}{g(x)}$ could assume any real value, $+\infty$, $-\infty$, or even have no limit. The choice of $f(x) = kx$, $g(x) = x$ and $c = 0$ provides such an example. There are seven indeterminate forms in all, and the best method for evaluating these is by the use of L'Hospital's Rule, as calculus students learn.

We usually think of the limit theorem in terms of the cases where L and M are finite, and $M \neq 0$ in the quotient case. However, the following chart shows that there are many more possibilities, and consequently many more cases to be proved to get a complete theorem. It also shows where the indeterminate forms are found. When we deal with continuous functions in Chapter 4, we will concentrate our attention on the first row, where L and M are both finite.

The following symbols occur at some place in the chart: $L + M$, LM, $\frac{L}{M}$, $+\infty$, $-\infty$, 0, $\pm\infty$, and I. The symbol $\pm\infty$ means that the function is unbounded, but doesn't approach $+\infty$ or $-\infty$. For example, $\lim_{x\to 0} \frac{1}{x} = \pm\infty$, since $\lim_{x\to 0-} \frac{1}{x} = -\infty$ and $\lim_{x\to 0+} \frac{1}{x} = +\infty$. The symbol I will denote an indeterminate form.

In each row, some minor changes in the proof are needed, depending on whether c is finite, $+\infty$, $-\infty$, or if a one-sided limit is taken, but the conclusions and main part of the proof will be unchanged. In some of the 27 boxes, there is more than one answer, depending on the values for L and M. For example, in row 2 and column 2, we must distinguish between the 3 cases where $L > 0$, $L = 0$, or $L < 0$.

Exercise 8 contains a result which is also part of a general limit theorem, but it is not included in the chart on page 86.

GENERAL LIMIT THEOREM FOR FUNCTIONS

Assume that $\lim_{x \to c} f(x) = L$ and $\lim_{x \to c} g(x) = M$

	$\lim_{x \to c} f(x) + g(x)$	$\lim_{x \to c} f(x)g(x)$	$\lim_{x \to c} \frac{f(x)}{g(x)}$
(1) L finite, M finite	$L + M$	LM	$\frac{L}{M}$ if $M \neq 0$ $\pm\infty$ if $M = 0,\ L \neq 0$ I if $M = 0,\ L = 0$
(2) L finite, $M = +\infty$	$+\infty$	$+\infty$ if $L > 0$ I if $L = 0$ $-\infty$ if $L < 0$	0
(3) L finite, $M = -\infty$			
(4) $L = +\infty$, M finite	$+\infty$	$+\infty$ if $M > 0$ I if $M = 0$ $-\infty$ if $M < 0$	$+\infty$ if $M > 0$ $\pm\infty$ if $M = 0$ $-\infty$ if $M < 0$
(5) $L = +\infty$, $M = +\infty$	$+\infty$	$+\infty$	I
(6) $L = +\infty$, $M = -\infty$			
(7) $L = -\infty$, M finite			
(8) $L = -\infty$, $M = +\infty$	I	$-\infty$	I
(9) $L = -\infty$, $M = -\infty$			

EXERCISES 11

1. Provide appropriate definitions for the remaining 3 cases in Definition 1.

2. In Definition 1, which form is used to define a convergent sequence? Which forms are used to express vertical asymptotes? Which forms are used to express horizontal asymptotes?

3. Use the appropriate form of Definition 1 in order to verify the following limits.

   (a) $\lim_{x \to 1^-} \frac{1}{x^2-1} = -\infty$

   (b) $\lim_{x \to -\infty} \frac{2}{x^3} = 0$

   (c) $\lim_{x \to 0^+} \frac{2}{x^3} = +\infty$

   (d) $\lim_{x \to 0^+} \log_{10} x = -\infty$

   (e) $\lim_{x \to \infty} \arctan x = \frac{\pi}{2}$

   (f) $\lim_{x \to \infty} x^3 = +\infty$

   (g) $\lim_{x \to c} \frac{1}{x} = \frac{1}{c}$

4. Prove that the form $\infty + \infty = \infty$ as follows: If $\lim_{x \to c} f(x) = +\infty$ and $\lim_{x \to c} g(x) = +\infty$, then $\lim_{x \to c} \left(f(x) + g(x) \right) = +\infty$.

5. Prove that the form $\infty(-\infty) = -\infty$ as follows: If $\lim f(x) = +\infty$ and $\lim g(x) = -\infty$, then $\lim f(x)g(x) = -\infty$.

6. Prove that the form $k \cdot \infty = +\infty$ if $k > 0$ as follows: If $\lim_{x \to c} f(x) = k > 0$ and $\lim_{x \to c} g(x) = +\infty$, then $\lim_{x \to c} f(x)g(x) = +\infty$.

7. Prove that if $\lim_{x \to c} f(x) = k > 0$, $\lim_{x \to c} g(x) = 0$, and $g(x) < 0$ for all $|x - c| < \delta$ for some value of δ, then $\lim_{x \to c} \frac{f(x)}{g(x)} = -\infty$.

8. Prove that if $\lim_{x \to c} f(x) = 0$, and $g(x)$ is bounded on some interval containing c, then $\lim_{x \to c} f(x)g(x) = 0$.

9. Fill in the remaining boxes in the chart on page 86 for the general limit theorem for functions.

SECTION 12 SEQUENCE AND NEIGHBORHOOD FORMS

After being possibly overwhelmed by all the cases in the limit definition of a function and in the Limit Theorem, you are likely to welcome any possible simplification. It turns out that the neighborhood concept from topology is able to do just that, permitting us to include all the previous cases in a single definition. Topology has become a very helpful tool to use when studying analysis. It abstracts and generalizes several concepts that occur repeatedly in different settings. This process of generalization is elegant and economical because it combines several cases into one. It also sheds light on the particular cases, since one can see them from a broader perspective. This is especially true for the topics of limits and continuous functions. On the other hand, topology helps very little in the units on the derivative, the integral, and infinite series because there is too much detailed material needed for those topics. We now see how the concept of the neighborhood of a point can be used to unify the several definitions of limits that we have studied.

Definition 1   The interval $(x - \epsilon, x + \epsilon)$ is an ϵ-*neighborhood of* x. The set U is a *neighborhood of* x if U contains an ϵ-neighborhood of x.

Example 1   The interval $(1, 2)$ is an ϵ-neighborhood of $\frac{3}{2}$, and is a neighborhood for any x for which $1 < x < 2$. The closed interval $[1, 2]$ is also a neighborhood for any x for which $1 < x < 2$, but it is not an ϵ-neighborhood for any x, and it is not a neighborhood of 1, even though it contains 1. $\square$

Definition 4.2 states that "the sequence $\{x_n\}$ converges to the real number x if for every $\epsilon > 0$, there is some N, so that $|x_n - x| < \epsilon$ for all $n > N$." Since the statement that $|x_n - x| < \epsilon$ is equivalent to the statement that $x_n \in (x - \epsilon, x + \epsilon)$, we could replace Definition 4.2 by "the sequence $\{x_n\}$ converges to x, if $\{x_n\}$ is eventually in every ϵ-neighborhood of x."

In a similar way, the symbol $\lim_{x \to c} f(x) = L$ was defined (see Definition 10.1) to mean that "for every $\epsilon > 0$, there is some $\delta > 0$ so that if $|x - c| < \delta$, then $|f(x) - L| < \epsilon$." This can be restated as $\lim_{x \to c} f(x) = L$

if "for every ϵ-neighborhood V of L, there is a δ-neighborhood U of c so that if $x \in U$, then $f(x) \in V$." The symbol $f(U) \subset V$ is often used to express the property that if $x \in U$, then $f(x) \in V$.

This same neighborhood terminology will cover the cases involving $\pm\infty$ if we make the following definition.

Definition 2   The open interval (c,∞) is a *neighborhood of $+\infty$* and the interval $(-\infty,c)$ is a *neighborhood* of $-\infty$.

Example 2   We restate three cases of limits in terms of the neighborhood concept as follows.

(1) The sequence $\{x_n\}$ converges to x if for every neighborhood V of x, there is a neighborhood U of $+\infty$, so that $x_n \in V$ for all $n \in U$.

(2) $\lim_{x \to c} f(x) = +\infty$ if for every neighborhood V of $+\infty$, there is a neighborhood U of c, so that $f(U) \subset V$. This is form (2) of Definition 11.1.

(3) $\lim_{x \to -\infty} f(x) = L$ if for every neighborhood V of L, there is a neighborhood U of $-\infty$, so that $f(U) \subset V$. This is form (7) of Definition 11.1.   □

THEOREM 12.1 (Neighborhood Form of Limit Definition)
Let $f : A \to \mathbf{R}$ and let c be a cluster point of A. Then $\lim_{x \to c} f(x) = L$ if and only if for every neighborhood V of L, there is a neighborhood U of c so that if $x \in A \cap U$, then $f(x) \in V$. We can also express this by $f(A \cap U) \subset V$.

Proof of $\Rightarrow$:   Assume that $\lim_{x \to c} f(x) = L$, and let V be a neighborhood of L. There is some $\epsilon > 0$ so that $(L - \epsilon, L + \epsilon) \subset V$ by Definition 1. By Definition 10.1, there is also some $\delta > 0$, so that if $x \in A$ and $|x - c| < \delta$, then $|f(x) - L| < \epsilon$. If we define $U = (c - \delta, c + \delta)$, then U is a neighborhood of c, and if $x \in A \cap U$, then $f(x) \in (L - \epsilon, L + \epsilon) \subset V$, which means that $f(A \cap U) \subset V$.   ∎

Proof of $\Leftarrow$:   Let $\epsilon > 0$ be given. This determines a neighborhood $V = (L - \epsilon, L + \epsilon)$ of L, so there is some neighborhood U of c, so that $f(A \cap U) \subset V$. By Definition 1, there is a $\delta > 0$, so that $(c - \delta, c + \delta) \subset U$. If $x \in A$ and $|x - c| < \delta$, then $x \in U$, and so $f(x) \in V$. This means that $L - \epsilon < f(x) < L + \epsilon$ implies $|f(x) - L| < \epsilon$ as desired.   ∎

Comment.   The main thing to observe is that if c or L takes on the "value" of $+\infty$ or $-\infty$, by use of Definition 2, we can still say that $\lim_{x \to c} f(x) = L$ is expressed by the statement that

> for every neighborhood V of L, there is a neighborhood U of c, so that $f(A \cap U) \subset V$.

This is illustrated by the next theorem, which shows that the definition of a convergent sequence can be expressed using the neighborhood formulation.

THEOREM 12.2   The sequence $\{x_n\}$ converges to x if and only if for every neighborhood V of x, there is a neighborhood U of $+\infty$, so that if $n \in U \cap \mathbf{N}$, then $x_n \in V$.

Proof of $\Rightarrow$:   If $\epsilon > 0$ is given, then $V = (x - \epsilon, x + \epsilon)$ is a neighborhood of x, so there is a neighborhood $U = (w, \infty)$ of $+\infty$, so that if $n \in U \cap \mathbf{N}$, then $x_n \in V$. Let $N = [w]$. If $n > N$, then since $n \in U \cap \mathbf{N}$, we have $x_n \in (x - \epsilon, x + \epsilon)$ or $|x_n - x| < \epsilon$.   ■

Proof of $\Leftarrow$:   This is left as an exercise.

We next prove a theorem that asserts the concept of a convergent sequence can be used to express the definition that $\lim_{x \to c} f(x) = L$. Therefore, we have three distinct ways to express the concept that a function f approaches the limit L as x approaches the value c. These are the ϵ–δ form, the neighborhood form, and the sequence form. The neighborhood form provides a bridge to topology, and gives an economical way to state definitions, although it is too general to be a useful device for proving many of our analysis theorems. The ϵ–δ form is historically the most widely used form, and every student of analysis should become proficient and comfortable with it. The sequence form has the advantage that it often permits us to prove a result more easily than with the use of epsilons and deltas, by using results about sequences that have been previously established.

Part of the proof of Theorem 12.3 requires the denial of the statement $\lim_{x \to c} f(x) = L$, so we next present that idea. It provides a model for negation, indicating that the phrases "for every" and "there exists" should be interchanged. To illustrate, in Definition 10.1, the symbol $\lim_{x \to c} f(x) = L$ is defined to mean that for each $\epsilon > 0$, there is a $\delta > 0$ with a certain property. This will fail if and only if there is some

$\epsilon_o > 0$ for which no δ has the desired property, which is that for all $x \in A$ so that $|x - c| < \delta$, it follows that $|f(x) - L| < \epsilon_o$. Failure of this property to hold implies that for each δ, there is some $x_\delta \in A$ so that $|x_\delta - c| < \delta$, but $|f(x_\delta) - L| \geq \epsilon_o$.

THEOREM 12.3 (Sequence Form of the Limit Definition)
Let $f : A \to \mathbf{R}$ and let c be a cluster point of A. Then $\lim_{x \to c} f(x) = L$ if and only if for every sequence $\{x_n\}$ so that $\lim_{n \to \infty} x_n = c$, and $x_n \in A$ for all n, we have $\lim_{n \to \infty} f(x_n) = L$.

Proof of $\Leftarrow$: Let $\epsilon > 0$ be given. There is some $\delta > 0$, so that if $x \in A$ and $|x - c| < \delta$, then $|f(x) - L| < \epsilon$. Let $\{x_n\}$ be a sequence, so that $x_n \in A$ for all n, and $\lim x_n = c$. There is some N_δ so that if $n > N_\delta$, then $|x_n - c| < \delta$, and so $|f(x_n) - L| < \epsilon$. Thus $\lim_{n \to \infty} f(x_n) = L$. ∎

Proof of $\Rightarrow$: Suppose that $\lim_{x \to c} f(x) \neq L$. There is some $\epsilon_o > 0$ so that for each $\delta > 0$, there is a x_δ so that $|x_\delta - c| < \delta$, but $|f(x_\delta) - L| \geq \epsilon_o$. In particular, for each n, there is some x_n so that $|x_n - c| < \frac{1}{n}$ but $|f(x_n) - L| \geq \epsilon_o$. This is a contradiction because $\{x_n\}$ is a sequence that converges to c, but $\{f(x_n)\}$ does not converge to L. ∎

EXERCISES 12

1. Let $f(x) = \sin \frac{1}{x}$. Prove that $\lim_{x \to 0+} f(x)$ does not exist by using the following steps.

   (a) First show that the sequence $\{x_n\} = \{\frac{1}{n\pi}\}$ is a sequence that converges to 0 from the right, such that $\{f(x_n)\}$ converges to 0.

   (b) Now find another sequence $\{y_n\}$ so that $\{y_n\}$ converges to 0 and $\{f(y_n)\}$ converges to 1.

   (c) Use Theorem 12.3 to complete this problem.

2. Prove that $\lim_{x \to 0+} \sin \frac{1}{x}$ does not exist by using a single sequence $\{x_n\}$ that converges to 0.

3. Prove that $\lim_{x \to 0+} \frac{1}{x}$ does not exist.

4. Prove the other half of Theorem 12.2.

5. Follow the pattern in Example 2 in order to restate forms (3), (4), and (6) of Definition 11.1 in terms of the neighborhood concept.

SECTION 13 TOPOLOGY FOR THE REAL LINE

In this section, we present information about the properties of three special categories of sets; namely open sets, closed sets, and compact sets.

Definition 1   Let $A \subseteq \mathbf{R}$. The point x is an *interior point of A* if there is a neighborhood U of x so that $U \subset A$. The point x is an *exterior point of A* if there is a neighborhood U of x so that $U \cap A = \emptyset$. A point x that is neither an interior point nor an exterior point of A is a *boundary point of A*.

Example 1   Let $a < c < b$ and let $k = \min \{c - a, b - c\}$. Then c is an interior point for the interval $A = (a, b)$ because $U = (c - k, c + k)$ is an ϵ-neighborhood of c that is contained in A. In a similar way, if $d < a$ or $d > b$, then d is an exterior point of A. Also, a and b are boundary points of A.   □

LEMMA 13.1   The point x is an interior point of the set A if and only if x is an exterior point of $\mathbf{R} \backslash A$. Recall that $\mathbf{R} \backslash A$ is the complement of the set A, and could also be expressed by $\mathcal{C}A$ (see Appendix 1).

Proof. x is an interior point of A
$\iff$ there is a neighborhood U of x so that $U \subset A$
$\iff$ there is a neighborhood U of x so that $U \cap \mathbf{R} \backslash A = \emptyset$
$\iff$ x is an exterior point of $\mathbf{R} \backslash A$ ∎

LEMMA 13.2   If x is an exterior point of the set A, then x is not a cluster point (or an accumulation point) of A.

Proof.   This is left as an exercise.

Definition 2   The set $A \subseteq \mathbf{R}$ is an *open set* if every point in A is an interior point of A.

Definition 3   The set $A \subseteq \mathbf{R}$ is a *closed set* if A contains all its cluster points.

Example 2   The intervals (a, b) and $(-\infty, a)$ are open sets and the intervals $[a, b]$ and $[a, \infty)$ are closed sets. The interval $[a, b)$ is neither open nor closed. It is not closed because it does not contain its cluster point b, and it is not open because a is not an interior point.   □

THEOREM 13.3   Let $A \subseteq \mathbf{R}$. Then A is open if and only if $\mathbf{R} \backslash A$ is closed.

Proof of $\Rightarrow$:   Let A be open and x be a cluster point of $\mathbf{R} \backslash A$. If x is in A, it must be an interior point of A, and thus an exterior point of $\mathbf{R} \backslash A$. This means by Lemma 13.2, that x is not a cluster point of $\mathbf{R} \backslash A$, which is a contradiction. Therefore x must be in $\mathbf{R} \backslash A$, and so $\mathbf{R} \backslash A$ is closed.   ■

Proof of $\Leftarrow$:   Let $\mathbf{R} \backslash A$ be a closed set and $x \in A$. This means that x cannot be a cluster point of $\mathbf{R} \backslash A$, so there must be a neighborhood U of x for which $U \cap \mathbf{R} \backslash A = \emptyset$, or equivalently, $U \subset A$. Therefore x is an interior point of A, and so A is open.   ■

We next consider the more difficult concept of compactness. To do this, we need to first discuss the idea of a collection of sets.

Definition 4   The sequence of sets $\{S_n\}_{n=1}^{\infty}$ is *increasing* if $S_n \subset S_{n+1}$ for all n, and *decreasing* if $S_{n+1} \subset S_n$ for all n.

Example 3   The sequence of closed intervals $\{[x - \frac{1}{n}, x + \frac{1}{n}]\}_{n=1}^{\infty}$ is decreasing, and $\cup_{n=1}^{\infty}[x - \frac{1}{n}, x + \frac{1}{n}] = [x - 1, x + 1]$. In addition, we have $\cap_{n=1}^{\infty}[x - \frac{1}{n}, x + \frac{1}{n}] = \{x\}$.   □

Example 4   The sequence $\{\mathbf{R} \backslash [x - \frac{1}{n}, x + \frac{1}{n}]\}$ is an increasing sequence of open sets with $\cup_{n=1}^{\infty} \mathbf{R} \backslash [x - \frac{1}{n}, x + \frac{1}{n}] = \mathbf{R} \backslash \{x\}$.   □

Definition 5   The collection T consisting of open sets is an *open cover for the set E* if every element in E is contained in some set G from T. We can write this as $E \subset \cup_{G \in T} G$. The set E has a *finite subcover from T* if there exists sets $G_1, G_2, \ldots G_n$ in T so that $E \subset \cup_{i=1}^{n} G_i$.

Example 5   The collection $T = \{(1 - \frac{1}{n}, 2 - \frac{1}{n}) : n \in \mathbf{N}\}$ is an open cover for the set $E = [\frac{1}{2}, 2)$, and the collection $T = \{(-n, n) : n \in \mathbf{N}\}$ is an open cover for the set $\mathbf{R} = (-\infty, \infty)$. These are both countable collections. Let $r \in \mathbf{R}$. Then $T = \{(r - 1, r + 1) : r \in \mathbf{R}\}$ is an uncountable collection of sets that is an open cover for the set $\mathbf{R}$.   □

Definition 6   The set $E \subseteq \mathbf{R}$ is a *compact set* if every open cover for E has a finite subcover.

Example 6   The set $[a, \infty)$ is not compact, since $T = \{(a - \frac{1}{n}, n)\}_{n=1}^{\infty}$ is an open cover for $E = [a, \infty)$, with the property that no finite subcollection covers E. Observe that $[a, \infty)$ is a closed set that is not bounded.   □

THEOREM 13.4   If E is a compact set, then E must be bounded.

Proof.   The collection $\{(-n, n) : n \in \mathbf{N}\}$ is an open cover for any set $E \subseteq \mathbf{R}$. The union of any finite subcollection will be an interval of the form $(-n_0, n_0)$. If E is compact, there must be an interval $(-n_0, n_0)$, so that $E \subset (-n_0, n_0)$, and so E is bounded.   ■

Example 7   The set $E = [a, b)$ is not compact, since the collection

$$T = \left\{(a - \frac{1}{n}, b - \frac{1}{n}) : n \in \mathbf{N}\right\}$$

is an open cover of E, but the union of any finite subcollection has the form $(a - \frac{1}{n_1}, b - \frac{1}{n_2})$, and does not therefore cover E. Observe that E is bounded but not closed.   □

THEOREM 13.5   If E is a compact set, then E must be closed. Equivalently, if E is not closed, then E is not compact.

Proof.   If E is not closed, there must be a cluster point x of E with $x \notin E$. The collection $\mathcal{T} = \{G_n\}_{n=1}^{\infty} = \{\mathbf{R}\backslash[x - \frac{1}{n}, x + \frac{1}{n}]\}_{n=1}^{\infty}$ (see Example 4) is an open cover for E. In order for E to be compact, there must be a finite subcollection $\{G_{n_i}\}_{i=1}^{k}$ of $\mathcal{T}$ that covers E. Since $\cup_{i=1}^{k} G_{n_i} = G_{n'}$ where $n' = \max\{n_1, n_2, \dots, n_k\}$, we have

$$E \subset \cup_{i=1}^{k} G_{n_i} = G_{n'} = \mathbf{R}\backslash[x - \frac{1}{n'}, x + \frac{1}{n'}]$$

or equivalently, $E \cap [x - \frac{1}{n'}, x + \frac{1}{n'}] = \emptyset$.

But $E \cap (x - \frac{1}{n'}, x + \frac{1}{n'}) = \emptyset$ implies that x is an exterior point of E, and therefore not a cluster point of E. This contradiction means that E is not compact, and the proof is complete.   ∎

Theorems 13.4 and 13.5 prove that a compact set must be both closed and bounded. The next theorem, called the Heine-Borel Theorem, asserts that these two conditions are also sufficient for compactness. It is a well-known result in analysis, and is equivalent to the nested interval property and the Bolzano-Weierstrass Theorem that were proved in Chapter 2. The Heine-Borel Theorem is an important result for a study of continuous functions.

THEOREM 13.6 (Heine-Borel Theorem)
The set $E \subseteq \mathbf{R}$ is compact if and only if E is closed and bounded.

Proof of $\Rightarrow$:   This half is proved in Theorems 13.4 and 13.5.

Proof of $\Leftarrow$:   We assume that E is a closed and bounded set, and prove that it is compact by a denial proof; that is, we assume the existence of an open cover $\mathcal{T}$ for E that has no finite subcover, and show that this assumption leads to a contradiction.

We begin by constructing a nest of closed intervals $\{I_n\}$ with the property that for any n, no finite collection of sets from $\mathcal{T}$ covers the set $E \cap I_n$. This is done by a process similar to the one used in the proof of the Bolzano-Weierstrass Theorem (Theorem 5.6). Since E is bounded, there must be an interval $I = [a, b]$ so that $E \subset I$. If I is bisected, then either the left half of I or the right half of I must have the property described above.

Let $I_1 = [a, \frac{a+b}{2}]$ if $I_1 \cap E$ cannot be covered by a finite subcollection
of sets from $\mathcal{T}$. Otherwise, let $I_1 = [\frac{a+b}{2}, b]$. Then define I_2 as the closed
left-half subinterval of I_1 if this half has the desired property, and as the
closed right-half subinterval of I_1 if a finite number of sets from $\mathcal{T}$ covers
the intersection of E and the left half of I_1. Continuing in this way, we
form a nest of closed intervals $\{I_n\}$ with the properties that for every n,
$\ell(I_n) = \frac{b-a}{2^n}$ and $I_n \cap E$ cannot be covered by a finite subcollection of sets
from $\mathcal{T}$. By the nested interval property (Theorem 5.5), there is a unique
number $x \in \cap I_n$.

Let $\epsilon > 0$ be given. There must be some I_n so that $I_n \subset (x - \epsilon, x + \epsilon)$
(see Exercise 5.19), and since $E \cap I_n$ is an infinite set for each n, it follows
that x is a cluster point of E. Since E was assumed to be closed, it
follows that $x \in E$. Since $\mathcal{T}$ is a cover for E, there must be some $G \in \mathcal{T}$
so that $x \in G$. Since G is open, there must be some $\epsilon_o > 0$ so that
$x \in (x - \epsilon_o, x + \epsilon_o) \subset G$, and by Exercise 5.19, there must also be some
I_{n_o}, so that $I_{n_o} \subset (x - \epsilon_o, x + \epsilon_o)$.

But then $E \cap I_{n_o} \subset I_{n_o} \subset G$, which says that a single set from $\mathcal{T}$ covers
$E \cap I_{n_o}$. This is the desired contradiction. Therefore, E is compact. ∎

EXERCISES 13

1. Prove Lemma 13.2.

2. Prove that x is a boundary point of A if and only if every neighborhood of x
   has the property that $U \cap A \neq \emptyset$ and $U \cap CA \neq \emptyset$.

3. Let $c < a$. Show that c is an exterior point of the set (a, ∞).

4. Prove that the interval $[a, \infty)$ is not compact by finding an open cover for
   $[a, \infty)$, so that no finite subcollection covers $[a, \infty)$.

5. Repeat Exercise 4 for the set $(a, b]$.

6. Prove Theorem 13.3 directly from Definitions 2 and 3 without the use of
   Lemmas 13.1 and 13.2.

7. Prove that if $\{S_n\}$ is a monotone increasing sequence of sets, then $\{\mathbf{R} \setminus S_n\}$ is
   a monotone decreasing sequence of sets.

8. Verify the assertions made in Example 4.

9. Prove that the interval $[a, b]$ is a closed set.

4

Continuity

SECTION 14 A HISTORICAL INTRODUCTION TO CONTINUITY

In the previous chapter, we developed general properties for the limit of a function, and are now ready to consider such categories of functions as continuous, differentiable, and integrable. We first pause to briefly discuss the evolution of the concept of a function. It is now customary to define a function by one of two approaches. The first approach is to use the idea of a mapping between two sets, where for any element x in the first set A, called the domain, the function f associates a single element y in the second set B, called the image space or codomain. While these two sets are usually subsets of the real numbers in this book, they may consist of other entities such as functions, complex numbers, or elements in Euclidean n-space. A second approach is to define a function from domain A to codomain B as a subset of the Cartesian product $A \times B$.

Both of these approaches are quite formal, compared with the early definitions of the function concept, which were strongly motivated by geometric or physical ideas. For example, while we today define a parabola as the graph of the function $f(x) = ax^2 + bx + c$, the Greeks defined it geometrically as an intersection of a plane with a right circular cone. In the 17th century, mathematicians might have defined the parabola in physical terms as the path of a thrown ball or fired projectile.

In the 17th century, many function-like entities, such as the logarithm, were introduced by such means as a verbal description, a table of values, or a graph. The stage was being set for a formal definition of a function through Viete's introduction in 1591 of symbols to represent operations and unknown quantities, and then through the application of this new algebra to geometry by Fermat and Descartes in the 1630s. Although Leibniz was the first to use the term "function" in his writings in the late 1600s, and while he also divided functions and curves into the two classes of algebraic and transcendental that we still observe, he did not envision a general concept of function.

From the time of Newton and Leibniz on through the entire 18th century, the usual way to represent the familiar functions was by means of infinite power series. This was later referred to as the analytic method of representation. Though the theory of convergence was not developed until the 19th century, these power series were used most effectively to represent functions and to obtain a host of useful results.

Euler was the first to attempt a definition of the function concept, being influenced by the ideas of his teacher John Bernoulli. In his famous textbook on analysis in 1748, the *Introductio in analysin infinitorum*, Euler proposed the definition that "a function of a variable quantity is an analytic expression composed in any way from this variable quantity and numbers or constant quantities." In another book in 1755, Euler broadened this definition to move closer to the modern notion of a correspondence between pairs of elements. For Euler, the term function meant continuous function as we think of it today, and these functions were also assumed to be differentiable and integrable throughout their domain. Since functions were considered as infinite polynomials at this time, it is no wonder that mathematicians thought there would be no problem in taking derivatives of all orders and integrating freely.

For example, if we start with the Maclaurin series

$$\sin x \; = \; x - \frac{x^3}{3!} + \frac{x^5}{5!} - \frac{x^7}{7!} + \cdots$$

and differentiate both sides, we obtain

$$\cos x \; = \; 1 - \frac{x^2}{2!} + \frac{x^4}{4!} - \frac{x^6}{6!} + \cdots$$

By integrating both sides of the series expression for $\sin x$, we find that

$$-\cos x \; = \; C + \frac{x^2}{2!} - \frac{x^4}{4!} + \frac{x^6}{6!} - \cdots$$

from which we show $C = -1$ by setting $x = 0$, and again get the familiar Maclaurin series for $\cos x$.

D'Alembert opposed Euler's concept of a function as being too carelessly defined, and Lagrange took turns supporting each of these views. Gradually, most mathematicians took Euler's side, and in the 1830s, the Russian mathematician Lobachevsky (who was also a co-founder of non-Euclidean geometry) and the German mathematician Dirichlet recast Euler's definition into its modern form. Here is Lobachevsky's definition—notice that it is still basically in essay form and has not yet assumed the language of sets, which would not be developed as a formal concept for another half century.

> General conception demands that a function of x be called a number which is given for each x and which changes gradually together with x. The value of the function could be given either by an analytical expression, or by a condition which offers a means for testing all numbers and selecting one of them; or lastly, the dependence may exist but remain unknown.[1]

Based on their experience in calculus, most students have the impression that continuity is a rather dry and unimportant topic, especially when compared with the interesting applications of the derivative and integral. In the historical development however, continuity was of crucial importance, being at the center of the transition from intuition to rigor that occurred during the first half of the nineteenth century.

Throughout the eighteenth century, continuity was an intuitive idea, often thought of in the following sense.

> If $x = f(t)$ is the distance x of a moving point from a fixed origin on a fixed straight line at time t, the function $f(t)$ possesses 'kinetic continuity' if the point in moving from the position $x_1 = f(t_1)$ at time t_1 to the position $x_2 = f(t_2)$ at a subsequent time t_2 passes through all the positions intermediate between x_1 and x_2.[2]

The Intermediate Value Theorem is a statement that this idea of kinetic continuity is implied by the modern definition. However, the function

$$f(x) = \begin{cases} x & \text{if } x \text{ is rational} \\ -x & \text{if } x \text{ is irrational} \end{cases}$$

has the intermediate value property on any closed interval containing 0, but is only continuous at 0. Bernard Bolzano provided a proof in 1817 of the result that "every number lying between two values of a continuous function is itself another value of that function," and then gave his definition that f is continuous at x if

---

[1] "The Concept of Function up to the Middle of the 19th Century" by A.P. Youschkevitch, *Archive for History of Exact Science*, 1, 1976, p. 77.

[2] *100 Years of Mathematics* by George Temple, London: Gerald Duckworth and Co. 1981, p. 136.

the difference $f(x + \omega) - f(x)$ can be made smaller than any given quantity, if one makes ω as small as one wishes.[3]

In 1821, Augustin Cauchy published his university lecture notes and introduced the modern idea of a limit based on small tolerances. In his definition of limit, Cauchy did not use the ϵ–δ notation that is so common today; that was introduced by Weierstrass around 1850.

For 50 years after this modern definition of continuity was available, virtually every mathematician believed that continuity and differentiability were identical concepts, except for possibly a few isolated points, such as $x = 0$ for the function $f(x) = |x|$. The possibility that a continuous function might fail to be differentiable for an infinite number of values was virtually unthinkable. Bolzano had given an example of such a function before 1830, but it was not discovered among his letters until after 1920.

In 1861, Weierstrass presented the function $f(x) = \sum_{n=0}^{\infty} a^n \cos(b^n \pi x)$ with appropriate conditions on a and b, and asserted that it was continuous for all x and yet did not have a derivative for even a single value of x. This example was soon known by most mathematicians and clearly showed that intuition could not be trusted to always provide correct results in calculus. It was because the usual functions studied in the calculus (i.e. algebraic, trigonometric, logarithmic, and exponential) were continuous for most values and had derivatives of all orders, that intuition served admirably for the first two centuries of development after the time of Newton and Leibniz. When Joseph Fourier applied calculus to the theory of heat in the opening years of the nineteenth century, new functions were involved that necessitated a careful distinction between continuity and differentiability. Example 2.3 shows that such distinctions are possible with familiar functions. A careful rigorous approach, such as the one used in this book, is the only safe way to proceed.

---

[3] *The Historical Development of the Calculus* by C. H. Edwards, New York: Springer-Verlag, 1979, p. 308.

SECTION 15   TYPES OF CONTINUOUS FUNCTIONS

We are now ready to present the definition of continuity, which is universally accepted, and which closely resembles the one given by Cauchy in 1821.

Definition 1   Let $f : I \to \mathbf{R}$ be given (i.e. f is a real-valued function defined on the interval $I \subset \mathbf{R}$) and let $c \in I$. f is *continuous at c* if $\lim_{x \to c} f(x) = f(c)$. According to Definition 10.1, this means that for each $\epsilon > 0$, there is some $\delta > 0$ so that if $x \in I$ and $|x - c| < \delta$, then $|f(x) - f(c)| < \epsilon$.

Definition 2   Let $f : I \to \mathbf{R}$ be given and let $B \subset I$. f is *continuous on the set B* if f is continuous at each $x \in B$.

Definition 3   Let $f : I \to \mathbf{R}$ be given and let $c \in I$. f is *right-continuous at c* if $\lim_{x \to c+} f(x) = f(c)$, and f is *left-continuous at c* if $\lim_{x \to c-} f(x) = f(c)$. Observe that f is continuous at c if and only if f is both right-continuous and left-continuous at c.

Example 1 $f(x) = \sqrt{x}$ is right-continuous at $c = 0$, continuous at all $c > 0$ and undefined for $c < 0$.   □

Example 2   Show that $f(x) = x^2$ is continuous at all $x \in \mathbf{R}$.

Solution.   Since $|f(x) - f(c)| = |x^2 - c^2| = |x + c| \cdot |x - c|$ and $|x + c| \leq |x| + |c| \leq 2|c| + 1$ if $\delta \leq 1$, Definition 1 is satisfied if we define $\delta = \min \left\{ 1, \frac{\epsilon}{2|c|+1} \right\}$.   □

Definition 4   Let $f : I \to \mathbf{R}$ be given and $c \in I$. f is *discontinuous at c* if f is not continuous at c. Four types of discontinuities are listed below.

   (a) f has an *infinite discontinuity at c* if either $\lim_{x \to c-} f(x)$ or $\lim_{x \to c+} f(x)$ is infinite.

(b) f has a *jump discontinuity* at c if $\lim_{x \to c^-} f(x)$ and $\lim_{x \to c^+} f(x)$ are finite but unequal.

(c) f has a *removable discontinuity* at c if $\lim_{x \to c} f(x)$ exists and is finite, but $f(c)$ is not defined to have the value of $\lim_{x \to c} f(x)$.

(d) f has an *oscillating discontinuity* at c if the function f is bounded and the one-sided limits do not exist.

Example 3   The function $f(x)$ as defined below has a removable discontinuity at $x = -1$ and infinite discontinuities at $x = 0$ and $x = 1$.

$$f(x) = \frac{x^2 + 3x + 2}{x^3 - x} = \frac{(x+1)(x+2)}{x(x-1)(x+1)}$$

We can remove the discontinuity at $x = -1$ by defining $f(-1) = \frac{1}{2}$.   □

Example 4   The function $f(x) = \sin \frac{1}{x}$ is continuous for all $x \neq 0$, and has an oscillating discontinuity at $x = 0$.   □

Example 5   Find the discontinuities of the function

$$f(x) = \begin{cases} \frac{1}{1-x} & \text{if } x < 0 \\ x^2 + 1 & \text{if } 0 \leq x \leq 2 \\ x + 4 & \text{if } x > 2 \end{cases}$$

Solution.   Because there are three separate parts to the definition of this function, Euler would have said that it is discontinuous at $x = 0$ and $x = 2$. By modern standards, since

$$\lim_{x \to 0^-} f(x) = \lim_{x \to 0^+} f(x) = f(0) = 1$$

the function is continuous at $x = 0$ (we will see in Chapter 5 that it is not differentiable there however). Since $\lim_{x \to 2^-} f(x) = 5$ and $\lim_{x \to 2^+} f(x) = 6$, f has a jump discontinuity at $x = 2$.   □

Example 6 $f(x) = [3x + 1]$ has a jump discontinuity whenever $3x + 1$ assumes the value of an integer n, which is for $x = \frac{n}{3}$ where $n \in \mathbf{Z}$.   □

Example 7   An example of a function that has a discontinuity at every real number is

$$f(x) = \begin{cases} 0 & \text{if } x \text{ is rational} \\ 1 & \text{if } x \text{ is irrational} \end{cases} \qquad \square$$

THEOREM 15.1 (Limit Theorem for Continuous Functions)
Let $f : I \to \mathbf{R}$ and $g : I \to \mathbf{R}$ be continuous at $c \in I$. Then the functions $f \pm g : I \to \mathbf{R}$ and $fg : I \to \mathbf{R}$ are continuous at c. If $g(c) \neq 0$, then $\frac{f}{g} : I \to \mathbf{R}$ is continuous at c.

Proof.   The proof follows from Theorem 10.2 and Definition 1.   ■

In addition to the ϵ–δ form of Definition 1 given above, the next theorem asserts that continuity could have been defined in terms of sequences, or in topological terms using neighborhoods. The sequence form is especially useful in showing functions to be discontinuous.

THEOREM 15.2   Let $f : I \to \mathbf{R}$ be defined and let $c \in I$. Then the following statements are equivalent.

(1) The function f is continuous at c (in the sense of Definition 1).

(2) For every sequence $\{x_n\}$ so that each $x_n \in I$ and $\lim_{n \to \infty} x_n = c$, we have $\lim_{n \to \infty} f(x_n) = f(c)$.

(3) For every neighborhood V of $f(c)$, there is a neighborhood U of c so that $f(I \cap U) \subset V$.

Proof of (1) $\Leftrightarrow$ (2):   This follows from Theorem 12.3.

Proof of (1) $\Leftrightarrow$ (3):   This follows from Theorem 12.1.   ■

Definition 5   Let the functions $f : A \to \mathbf{R}$ and $g : B \to \mathbf{R}$ be defined so that $f(A) \subset B$. Let $c \in A$ and $d = f(c)$. The *composite function* $g \circ f$ is defined on domain A by the formula $(g \circ f)(c) = g(d)$.

Comment.   You should observe that $f \circ g$ is usually a very different function than $g \circ f$. In fact, $f \circ g$ may not even be defined. For example, if $f(x) = \ln x$ and $g(x) = -e^x$, then $(g \circ f)(x) = -x$ for all $x > 0$, but $(f \circ g)(x)$ is not defined for any value of x. When g is the inverse function f^{-1}, we have $f \circ f^{-1} = f^{-1} \circ f = i$ (the identity function). In Chapter 5, the Chain Rule provides a formula for differentiating a composite function.

Example 8   Let $f(x) = \ln x$, $g(x) = \sqrt{x}$, and $h(x) = e^x$. The function $(g \circ h)(x) = \sqrt{e^x}$ is defined for all x, while $(h \circ g)(x) = e^{\sqrt{x}}$ is defined for all $x \geq 0$, and the two functions give the same value only for $x = 0$ and $x = 4$. Also $(f \circ h)(x) = x$ for all x, while $(h \circ f)(x) = x$ for $x > 0$, so $f = h^{-1}$ on the domain of $x > 0$.   □

THEOREM 15.3   Let $f : A \rightarrow \mathbf{R}$ and $g : B \rightarrow \mathbf{R}$ be defined so that $f(A) \subset B$. If f is continuous at $c \in A$ and g is continuous at $d = f(c)$, then $g \circ f$ is continuous at c.

Proof.   Given $\epsilon > 0$, we must find δ so that if $x \in A$ and $|x - c| < \delta$, then $|(g \circ f)(x) - (g \circ f)(c)| < \epsilon$. Since g is continuous at d, there is $\gamma(\epsilon) > 0$ so that $|g(y) - g(d)| < \epsilon$ whenever $y \in B$ and $|y - d| < \gamma(\epsilon)$. Notice that this can be restated as $|g(f(x)) - g(f(c))| = |(g \circ f)(x) - (g \circ f)(c)| < \epsilon$ if $|f(x) - f(c)| < \gamma(\epsilon)$. But since f is continuous at c, there is $\delta(\gamma)$ so that if $x \in A$ and $|x - c| < \delta(\gamma)$, then $|f(x) - f(c)| < \gamma(\epsilon)$. By setting $\delta = \delta(\gamma)$, the proof is complete.   ∎

Example 9 $f(x) = \ln x$ is continuous for all $x > 0$, while $g(x) = \sin x$ is continuous for all $x \in \mathbf{R}$. Then $(g \circ f)(x) = \sin(\ln x)$ is continuous for all $x > 0$ and undefined for $x \leq 0$. $(f \circ g)(x) = \ln(\sin x)$ is continuous for all x for which $\sin x > 0$, which is on any interval of the form $(2n\pi, 2n\pi + \pi)$, where $n \in \mathbf{Z}$.   □

EXERCISES 15

1. Find all the discontinuities for each of the following functions, and identify each one as removable, infinite, jump or oscillating.

   (a) $f(x) = \frac{x^2 + 2x}{x^2 - 4}$

   (b) $f(x) = [x + \frac{1}{2}]$

   (c) $f(x) = [\frac{1}{x}]$

(d) $f(x) = \log_b x$

(e) $f(x) = \csc x$

(f) $f(x) = \frac{\sin x}{x}$

(g) $f(x) = \frac{2^x - 1}{x}$

(h) $f(x) = x \sin \frac{1}{x}$

(i) $f(x) = \cos \frac{1}{x}$

(j) $f(x) = \frac{\sqrt{x}-2}{x-4}$

(k) $f'(x)$, where $f(x) = |x|$

(l) $f(x) = 0$, if x is rational and $f(x) = 1$ if x is irrational.

(m) $f(x) = \frac{x^2 + 2x}{x^3 - x^2}$

2. Find an expression for δ which satisfies Definition 1 for the following functions at a given point c.

(a) $f(x) = \sqrt{x}$

(b) $f(x) = \frac{1}{x+2}$

(c) $f(x) = x^3$

3. (a) Give a definition for the composite function $f \circ g$.

(b) Provide an example of functions f and g so that $f \circ g \neq g \circ f$.

4. Prove that if the sequence $\{x_n\}$ converges to x, and f is continuous on an interval containing x, then the sequence $\{f(x_n)\}$ converges to $f(x)$.
Use this result to give an alternate solution for Example 4.5.

5. Prove Theorem 15.1 for fg, using Theorem 10.1.

6. Show that the discontinuity of $f(x) = \frac{1}{x} \sin \frac{1}{x}$ does not fit any of the 4 categories in Definition 4. How might the categories by redefined so as to handle this type of discontinuity?

7. Prove that if the functions $f_1, f_2, \ldots, f_n$ are all continuous at $c \in I$, then the function $f = \sum_{i=1}^{n} f_i$ will also be continuous at $c \in I$.

8. Use part (2) of Theorem 15.2 to prove that

(a) the function $f(x) = \frac{3}{x-2}$ is discontinuous at $x = 2$.

(b) the function $f(x) = [ax + b]$ is discontinuous at $x = \frac{2-b}{a}$.

SECTION 16 PROPERTIES OF CONTINUOUS FUNCTIONS

In this section, we prove some important properties for functions that are continuous on a closed interval. These three theorems are now stated as a reference for the comments about them that follow. Each theorem will then be restated later just before its proof.

THEOREM 16.1 (Boundedness of Continuous Functions)
If $f : I \to \mathbf{R}$ is continuous on the closed interval $I = [a, b]$, then f is bounded on I.

THEOREM 16.2 (Extreme Value Theorem)
Let $f : I \to \mathbf{R}$ be continuous on the closed interval $I = [a, b]$ and let $m = \inf \{f(x) : x \in I\}$ and $M = \sup \{f(x) : x \in I\}$. Then there must be values c and d in I so that $f(c) = m$ and $f(d) = M$.

THEOREM 16.3 (Intermediate Value Theorem)
Let $f : I \to \mathbf{R}$ be continuous on the closed interval $I = [a, b]$ and let $f(a) \neq f(b)$. For any value t between $f(a)$ and $f(b)$, there is a value $c \in (a, b)$ so that $f(c) = t$.

Most of these results can be proved by use of the Bolzano-Weierstrass Theorem for sequences. Theorem 16.2 is used in Chapter 5 to prove the Mean Value Theorem. The main use of Theorem 16.3 is to prove some results about the existence of zeroes of functions.

All of these theorems can be proved only if the given function f is known to be continuous on a closed interval. For example, with reference to Theorem 16.1, $f(x) = \frac{1}{x}$ is continuous on the open interval $(0, 1)$, but it is not bounded there. The underlying reason why the interval must be closed is a topological one; it is necessary that the set be compact, which means that it must be bounded and closed (see Theorem 13.6). As we prove these theorems, you should observe the important role that cluster points of sets play in this development.

In Lemma 4.2, we proved that a convergent sequence is bounded, using only the definition of convergence, and we obtained a bound of the form $B = \max \{|x_1|, |x_2|, \ldots, |x_N|, |x| + 1\}$. In Lemma 8.1, we proved a similar result for a Cauchy sequence. We now want to prove a boundedness result for a continuous function on a closed interval. From the definition of continuity, we know that if f is continuous at $x = c$, then f is bounded over some interval containing c. For example, $|f(x)| \le |f(c)| + 1$ for all x for which $|x - c| < \delta_1$. This result might suggest the following type of direct proof for Theorem 16.1.

An incorrect proof for Theorem 16.1

(1) Since f is continuous at $x = a$, there is some interval about a, call it (a_1, b_1), where f is bounded, say by $B_1 = |f(a)| + 1$.

(2) Since f is continuous at b_1, there is some interval about b_1, call it (a_2, b_2), where f is bounded, say by B_2. See Figure 16.1.

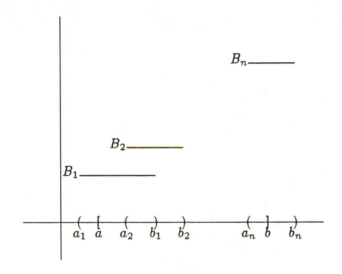

Figure 16.1

(3) Continuing in this manner, we move across the interval $[a, b]$ until we reach the interval (a_n, b_n) which contains the point b. A bound for f on $[a, b]$ is $B = \max\{B_1, B_2, \ldots, B_n\}$. ∎

This proof is very similar in style to the one we gave for Lemmas 4.2 and 8.1, and you may be hard pressed to find anything wrong with it. A proof by Cauchy that has a similar problem will be discussed in Section 30. Example 1 provides a hint of the problem with this proposed proof.

Example 1   The function $f(x) = \frac{1}{x}$ is continuous on $(0, 1]$. Since f is continuous at 1, f is bounded by $B_1 = 2$ on the interval $(\frac{1}{2}, 1)$. Since f is continuous at $\frac{1}{2}$, f is bounded by $B_2 = 3$ on the interval $(\frac{1}{3}, \frac{1}{2})$. Continuing in this way, since f is continuous at $x = \frac{1}{n}$, f is bounded by $B_n = n + 1$ on the interval $(\frac{1}{n+1}, \frac{1}{n})$ (see Figure 16.2). It should be clear that at any stage, $B = \max\{B_1, B_2, \ldots, B_n\}$ will not serve as a bound for f on the entire interval $(0, 1]$. $\square$

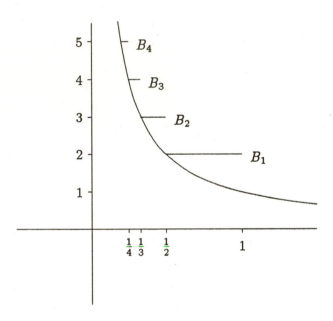

Figure 16.2

We are now ready to present an acceptable proof for Theorem 16.1. We will have to use a denial proof, and appeal to the Bolzano-Weierstrass Theorem.

THEOREM 16.1 (Boundedness of Continuous Functions)
If f is continuous on the closed interval $I = [a, b]$, then f is bounded on I.

Proof.   If we deny the conclusion that f is bounded on $[a, b]$, we need to express what it means for f to be unbounded on a set. If we use the principle suggested in Section 12, of interchanging "there exists" statements and "for every" statements, *unboundedness* would mean that for every bound we might choose, f exceeds this bound for at least one value in $[a, b]$. We can then form a sequence, if for every positive integer n, we let x_n represent a point in $[a, b]$, so that $|f(x_n)| > n$. $\{x_n\}$ is then a bounded sequence in $[a, b]$, while $\{|f(x_n)|\}$ is an unbounded sequence that diverges to $+\infty$.

By Theorem 7.1, $\{x_n\}$ has a convergent subsequence, which we denote by saying that $\{x_{n_k}\}$ converges to x_o. Since all the x_{n_k} values are in $[a, b]$, then x_o must also be in $[a, b]$, since this closed set contains all of its cluster points. Because f is continuous at x_o we use Theorem 15.2 to claim that $\lim_{k \to \infty} f(x_{n_k}) = f(x_o)$. But this is a contradiction to the earlier choice that $|f(x_{n_k})| > n_k$, which implies that $\lim_{k \to \infty} |f(x_{n_k})| = +\infty$.

Since a contradiction has been reached by denying the boundedness of f on $[a, b]$, we conclude that f must be bounded on $[a, b]$ and the proof is complete. This proof may appear to be longer than it is because of all the explanatory comments. You may wish to redo the proof, including only the essential steps.   ∎

We next come to Theorem 16.2, and provide two different proofs for this result.

THEOREM 16.2 (Extreme Value Theorem)
Let f be continuous on the closed interval $I = [a, b]$ and let

$$m = \inf\{f(x) : x \in I\} \quad \text{and} \quad M = \sup\{f(x) : x \in I\}$$

Then there must be values c and d in I so that $f(c) = m$ and $f(d) = M$.

First proof.   This is a proof by contradiction that uses Theorem 16.1. If we claim that there is no value d in $[a, b]$ for which $f(d) = M$, then it must be that $f(x) < M$ for all $x \in [a, b]$. But

$$g(x) = \frac{1}{M - f(x)}$$

is continuous on $[a, b]$ by Theorem 15.1, and is bounded on $[a, b]$ by Theorem 16.1. Suppose that $|g(x)| \leq B$, for all $x \in [a, b]$. Since

$$g(x) = \frac{1}{M - f(x)} \leq B \quad \text{is equivalent to} \quad f(x) \leq M - \frac{1}{B}$$

we have a contradiction to the choice of M as the least upper bound of f on $[a, b]$.   ∎

Second proof. This is a direct proof and makes use of the Bolzano-Weierstrass Theorem for sequences, in a way similar to the proof of Theorem 16.1. This time we present the argument without explanatory comments. You should supply these as needed.

Using the definition of $M = \sup\{f(x) : a \le x \le b\}$, for each n, there is some $x_n \in [a,b]$ so that $M - \frac{1}{n} < f(x_n) \le M$, and therefore $\lim_{n\to\infty} f(x_n) = M$. Now $\{x_n\}$ must have a convergent subsequence $\{x_{n_k}\}$, which converges to some $d \in [a,b]$. By the continuity of f at d, we have $\lim_{k\to\infty} f(x_{n_k}) = f(d)$. Since $\{f(x_n)\}$ converges to M, the subsequence $\{f(x_{n_k})\}$ must converge to M, and therefore $f(d) = M$.

Similarly, it can be shown there is some $c \in [a,b]$, so that $f(c) = m$. ∎

THEOREM 16.3 (Intermediate Value Theorem)

Let f be continuous on the closed interval $I = [a,b]$ so that $f(a) \ne f(b)$, and let t be any value between $f(a)$ and $f(b)$. Then there must be some value $c \in I$ so that $f(c) = t$.

Proof. Assume that $f(a) < t < f(b)$. Let $x_1 = \frac{a+b}{2}$ be the midpoint of the interval $I = [a,b]$. If $f(x_1) = t$, the proof is complete. If $f(x_1) > t$, let $I_1 = [a_1, b_1]$ be the left-half subinterval of I. If $f(x_1) < t$, let $I_1 = [a_1, b_1]$ be the right-half subinterval of I. Observe that $f(a_1) < t < f(b_1)$.

Let x_2 be the midpoint of I_1. If $f(x_2) = t$, the proof is complete. If $f(x_2) > t$, let I_2 be the left-half subinterval of I_1, and if $f(x_2) < t$, let I_2 be the right-half subinterval of I_1. Observe that if $I_2 = [a_2, b_2]$, then $f(a_2) < t < f(b_2)$.

Continuing in this way, we construct the nest of closed intervals $\{I_n\}$ so that if $I_n = [a_n, b_n]$, then $\lim_{n\to\infty} a_n = x = \lim_{n\to\infty} b_n$. By using Theorem 15.2,

$$f(x) = f\left(\lim_{n\to\infty} a_n\right) = \lim_{n\to\infty} f(a_n) \quad \text{and} \quad f(x) = f\left(\lim_{n\to\infty} b_n\right) = \lim_{n\to\infty} f(b_n)$$

Also, $\lim_{n\to\infty} f(a_n) \le t \le \lim_{n\to\infty} f(b_n)$ by Theorem 4.6. Putting all these results together, we obtain $f(x) = t$, which is the desired result.

The case where $f(a) > f(b)$ can be handled by applying the above result to the function $g(x) = -f(x)$. ∎

COROLLARY 16.4 (Location of Roots)

If f is continuous on the closed interval $[a, b]$, and $f(a)f(b) < 0$, then there is some point $c \in (a, b)$, so that $f(c) = 0$.

Proof. This is left as an exercise.

There is one more interesting feature about the statements of Theorems 16.2 and 16.3. It appears that these results could be combined into a single theorem, which asserts that if f is continuous on $[a, b]$, then for any y so that $m \leq y \leq M$, there is some $x \in [a, b]$, so that $f(x) = y$. However, the proof form that is used when y assumes an extreme value of m or M is significantly different from the proof form used when y assumes a strictly intermediate value $m < y < M$. This is one reason why these results are stated as two separate theorems. Also, the consequences that follow from each of these theorems are very different.

EXERCISES 16

1. Illustrate the Extreme Value Theorem for $f(x) = \frac{1}{3}x^3 - x^2 - 3x + 4$ on the interval $[0, 5]$. Begin by finding the values for m and M.

2. Use the Intermediate Value Theorem to show that the polynomial $p(x) = x^4 + 7x^3 - 9$ must have 2 real zeroes. You may use $p'(x)$ to help you with this problem.

3. Give an example of a function f that is defined on a closed interval, but is discontinuous and unbounded there. This example illustrates that the continuity part of Theorem 16.1 is essential.

4. Give an example of a continuous function f on an open interval (a, b) for which the conclusion of the Extreme Value Theorem does not hold. Give another example of a continuous function f on (a, b) for which the conclusion of the Extreme Value Theorem does hold.

5. For which values of y will it be true that there is a value $x \in [0, 5]$, so that $y = x^3 - 6x^2 + 9x + 1$? Which theorem or theorems would you use to substantiate your answer?

6. Prove Corollary 16.4.

7. Use Corollary 16.4 to prove there is some x so that $0 < x < \frac{\pi}{2}$ and $\cos x = x$.

8. Provide a proof of Theorem 16.2 for the case showing the existence of a value $c \in [a, b]$, so that $f(c) = m$.

9. What is the problem with the incorrect proof suggested for Theorem 16.1?

10. Write the proof for Theorem 16.1 in the step-reason format.

SECTION 17   UNIFORM CONTINUITY

We have shown that $f(x) = \frac{1}{x}$ is continuous for all $c \neq 0$, and found that $\delta_\epsilon = \min\{\frac{|c|}{2}, \frac{\epsilon c^2}{2}\}$ satisfies the limit definition for continuity. In actual fact, δ really depends on the value of c as well as the value of ϵ, so we could write $\delta(\epsilon, c) = \delta_{\epsilon,c} = \min\{\frac{|c|}{2}, \frac{\epsilon c^2}{2}\}$. For some sets S, it is possible to find a single value for δ that works "uniformly" for all $c \in S$. If $\inf\{\delta_{\epsilon,c} : c \in S\} > 0$, we can let this value serve as the "uniform δ."

We are now ready to express a formal definition for uniform continuity on a set S. Remember the intuitive idea is that there is a single value for δ_ϵ, which satisfies the ϵ–δ continuity definition for all $x \in S$.

Definition 1   The function f is *uniformly continuous on the set S* if for every $\epsilon > 0$, there is a $\delta > 0$ so that for all $x, u \in S$ so that $|x - u| < \delta$, we have $|f(x) - f(u)| < \epsilon$.

Example 1   For $f(x) = \frac{1}{x}$ on $S = [1, 3]$, the values for δ vary between $\frac{\epsilon}{2}$ and $\frac{9\epsilon}{2}$. So $\delta_\epsilon = \frac{\epsilon}{2}$ is a uniform continuity δ for the set $S = [1, 3]$. In a similar way, $\delta_\epsilon = 2\epsilon$ is a uniform δ for the set $S = [2, 5]$, $\delta_\epsilon = \frac{\epsilon}{2}$ is a uniform δ for the set $S = [1, \infty)$, and $\delta_\epsilon = \frac{\epsilon}{200}$ is a uniform δ for the set $S = [\frac{1}{10}, 2]$. For this function, we observe that the minimum value for δ occurs at the left endpoint of the interval. For the set $S = (0, 1]$, since

$$\inf\{\delta_{\epsilon,c} : c \in (0, 1]\} \;=\; \inf\left\{\frac{\epsilon c^2}{2} : c \in (0, 1]\right\} \;=\; 0$$

there is apparently no single value of δ that works uniformly on the whole set $(0, 1]$. This in itself will not disprove uniform continuity, but it suggests that possibility.   □

Example 2 $f(x) = \frac{1}{x}$ is uniformly continuous on the set $S = [1, 3]$ because

$$|f(x) - f(u)| \;=\; \left|\frac{1}{x} - \frac{1}{u}\right| \;=\; \left|\frac{u - x}{xu}\right| \;=\; \frac{1}{|x|} \cdot \frac{1}{|u|} \cdot |x - u| \;<\; 1 \cdot 1 \cdot \delta$$

since for $x, u \in [1, 3]$, we have $\frac{1}{3} \leq \frac{1}{x} \leq 1$ and $\frac{1}{3} \leq \frac{1}{u} \leq 1$. Therefore $\delta = \epsilon$ is a delta of uniform continuity.   □

Since the proof of Theorem 17.2 is done by denial, we present next a lemma to express what it means for a function not to be uniformly continuous on a set S. It is useful to present this idea in terms of sequences, rather than in the denial form of Definition 1.

LEMMA 17.1 The function f is not uniformly continuous on the set S if there is some $\epsilon_o > 0$ and a pair of sequences $\{x_n\}$ and $\{u_n\}$ in S, so that $\lim_{n\to\infty}(x_n - u_n) = 0$ and $|f(x_n) - f(u_n)| \geq \epsilon_o$.

Proof. This is left as an exercise.

In order to verify that a given function f is not uniformly continuous on a set S, we need to identify the part of the set at which the problem occurs, and then choose appropriate sequences in this region.

Example 3 $f(x) = \frac{1}{x}$ is not uniformly continuous on $S = (0, 1]$ because the function is unbounded near zero.

Solution. To prove this by Lemma 17.1, let $\epsilon_o = 1$, $\{x_n\} = \{\frac{1}{n}\}$, and $\{u_n\} = \{\frac{1}{2n}\}$. Then

$$\lim_{n\to\infty}(x_n - u_n) = \lim_{n\to\infty}\left(\frac{1}{n} - \frac{1}{2n}\right) = \lim_{n\to\infty}\frac{1}{2n} = 0$$

but $|f(x_n) - f(u_n)| = |n - 2n| = n \geq 1$. $\square$

Example 4 $f(x) = x^2$ is not uniformly continuous on $S = [1, \infty)$, because of the unboundedness of the function as x approaches $+\infty$. We accordingly can set $\{x_n\} = \{n\}$ and $\{u_n\} = \{n + \frac{1}{n}\}$, so that

$$|f(x_n) - f(u_n)| = \left(n + \frac{1}{n}\right)^2 - n^2 = 2 + \frac{1}{n^2} > 2 \quad \square$$

THEOREM 17.2 If f is continuous on the closed interval $I = [a, b]$, then f is uniformly continuous on I.

Proof. If we deny the conclusion, there is some $\epsilon_o > 0$ and two sequences $\{x_n\}$ and $\{u_n\}$, so that $\lim_{n\to\infty}(x_n - u_n) = 0$ and $|f(x_n) - f(u_n)| \geq \epsilon_o$. Since $\{x_n\}$ is a bounded sequence, there must be a convergent subsequence, which we denote by saying that $\{x_{n_k}\}$ converges to x_o. But then the corresponding subsequence $\{u_{n_k}\}$ must also converge to x_o because of the inequality

$$|u_{n_k} - x_o| \leq |u_{n_k} - x_{n_k}| + |x_{n_k} - x_o| < \frac{\epsilon}{2} + \frac{\epsilon}{2} = \epsilon$$

But since the sequences $\{x_{n_k}\}$ and $\{u_{n_k}\}$ both converge to x_o, and f is continuous at x_o, it follows that the sequences $\{f(x_{n_k})\}$ and $\{f(u_{n_k})\}$ must both converge to $f(x_o)$ or $\lim_{n \to \infty} |f(x_{n_k}) - f(u_{n_k})| = 0$, which is a blatant contradiction to the assumption that $|f(x_{n_k}) - f(u_{n_k})| \geq \epsilon_o$ for all k. ∎

Example 5 By Theorem 17.2, $f(x) = \frac{1}{x}$ is uniformly continuous on any closed interval that does not contain 0. By Definition 1, f will also be uniformly continuous on sets of the form $[c, \infty)$ for $c > 0$, because if $x, u \in [c, \infty)$ and $|x - u| < \delta = c^2\epsilon$, then

$$\left| \frac{1}{x} - \frac{1}{u} \right| = \left| \frac{u - x}{xu} \right| = \frac{1}{|x|} \cdot \frac{1}{|u|} \cdot |x - u| < \frac{1}{c} \cdot \frac{1}{c} \cdot c^2\epsilon = \epsilon \quad \square$$

The concept of uniform continuity is a theoretical idea that is important for the proofs of other theorems. In this regard, it is similar to the concept of a Cauchy sequence. The main theorem we will prove using Theorem 17.2 is that all continuous functions are integrable (see Section 26). It would also be helpful to obtain additional results like the one in Theorem 17.2 that guarantee uniform continuity on sets other than closed intervals, or provide conditions under which a function is not uniformly continuous. The Mean Value Theorem in Chapter 5 will enable us to prove such a result.

After a new concept has been defined, it is a good learning experience for you to try to formulate valid results connecting the new concept with previous concepts and results. Intuition and guesswork are important tools at this point. Proposed theorems should be checked by seeing if they hold for several specific examples. If a counterexample is found by this process, of course the result cannot be true, but it may be that the statement of the result could be altered somewhat to avoid the problem with the given example. This is the method proposed by Imre Lakatos as an important means of mathematical discovery in his book *Proofs and Refutations*, which was published by Cambridge University Press in 1976.

You are now given the opportunity to practice this approach. Several possible conjectures about uniform continuity are listed below, all of which should seem intuitively plausible. In the exercises, you are asked to decide which ones are true, and to find counterexamples for those that are false. It is usually helpful to consider different kinds of sets. For example, if S is a closed interval, then Conjecture 1 is a restatement of Theorem 17.2, but Conjecture 1 is false for some other types of sets.

Conjecture 1 If f is bounded and continuous on S, then f is uniformly continuous on S.

Conjecture 2 If f is unbounded on S, then f is not uniformly continuous on S.

Conjecture 3 If f has an unbounded derivative on S, then f is not uniformly continuous on S.

Conjecture 4 If f has a bounded derivative on S, then f is uniformly continuous on S.

Conjecture 5 If f is unbounded and f' is unbounded on S, then f is not uniformly continuous on S.

EXERCISE 17

1. Find a δ of uniform continuity that satisfies Definition 1 for the following functions on the given set S.
   (a) $f(x) = \frac{1}{x^2}$ on $S = [c, \infty)$, where $c > 0$
   (b) $f(x) = x^3$ on $S = [0, b]$
   (c) $f(x) = \sqrt{x}$ on $S = [c, \infty)$, where $c > 0$
   (d) $f(x) = \frac{1}{x-1}$ on $S = (-\infty, c)$, where $c < 1$

2. (a) Show that $\delta = \epsilon\sqrt{c}$ is a δ of continuity for $f(x) = \sqrt{x}$ at $x = c > 0$.
   (b) Since $\inf\{\epsilon\sqrt{c} : 0 \le c \le 1\} = 0$, it would seem that there is no δ of uniform continuity for $f(x) = \sqrt{x}$ on $[0, 1]$. However, we know that $f(x) = \sqrt{x}$ is uniformly continuous on $[0, 1]$. Why?
   (c) Can you explain this apparent contradiction?

3. Use Lemma 17.1 to show that each function below is not uniformly continuous on the given set S.
   (a) $f(x) = \frac{1}{x^2}$ on $S = (0, 1)$

(b) $f(x) = \frac{1}{x-1}$ on $S = (1, \infty)$

(c) $f(x) = \sin \frac{1}{x}$ on $S = (0, 1)$

(d) $f(x) = x \sin x$ on $S = [0, \infty)$

4. On which sets will $f(x) = \begin{cases} x \sin \frac{1}{x} & \text{if } x \neq 0 \\ 0 & \text{if } x = 0 \end{cases}$ be uniformly continuous?

5. Prove that $f(x) = \frac{1}{1+x^2}$ is uniformly continuous on the set of all reals $\mathbf{R} = (-\infty, +\infty)$.

6. Prove Lemma 17.1.

7. (a) Find an unbounded function f that is uniformly continuous on the set $[0, \infty)$.

   (b) Find a function f that has an unbounded derivative, but is uniformly continuous on $[0, 1]$.

8. Prove that if f is continuous on (a, b) and has an infinite discontinuity at a, then f is not uniformly continuous on (a, b).

Carry out the following process for the conjectures stated in Exercises 9-13. In each of the three cases where

(1) S is a closed interval $[a, b]$

(2) S is an open interval (a, b)

(3) S is an infinite interval $[a, \infty)$

state whether you believe each result is true or false. If you believe it is false, supply a counterexample.

9. Conjecture 1 – If f is bounded and continuous on S, then f is uniformly continuous on S.

10. Conjecture 2 – If f is unbounded on S, then f is not uniformly continuous on S.

11. Conjecture 3 – If f has an unbounded derivative on S, then f is not uniformly continuous on S.

12. Conjecture 4 – If f has a bounded derivative on S, then f is uniformly continuous on S.

13. Conjecture 5 – If f is unbounded and f' is unbounded on S, then f is not uniformly continuous on S.

14. Provide a proof for one of the cases in Exercises 9–13 that you believe is true.

15. Let $f : A \to \mathbf{R}$ be uniformly continuous, and $\{x_n\}$ be a Cauchy sequence in A. Prove that $\{f(x_n)\}$ is a Cauchy sequence in the set $\mathbf{R}$.

16. Exercise 3d shows that it is not generally true that if f and g are uniformly continuous on $\mathbf{R}$, then the product fg is uniformly continuous on $\mathbf{R}$. Prove this result is true however if f and g are also bounded in $\mathbf{R}$.

5

Differentiation

SECTION 18 A HISTORICAL INTRODUCTION TO DIFFERENTIATION

A common statement often made in a calculus course is that Newton and Leibniz discovered calculus, but the student probably has little idea of what is really meant by this statement. It is certainly not true that Newton and Leibniz were the first to find slopes of tangent lines to curves or areas under curves, since many individuals preceded them in these accomplishments. In fact, Archimedes had solved some fairly complicated area and volume problems almost two thousand years earlier. However, two things were lacking in the efforts of the predecessors of Newton and Leibniz. The first was the perception that what we now call differentiation and integration are general processes that can be applied widely, rather than specialized techniques for solving specific problems. The second deficiency was the lack of understanding that the problems of tangents and quadratures (as areas used to be called) were inverse in nature. Nevertheless, Pierre Fermat, Blaise Pascal, Isaac Barrow, John Wallis, and others played an essential role as forerunners of the discovery of the calculus. As Newton put it, "If I have seen further than others, it is because I have stood on the shoulders of giants." Both Newton and Leibniz cited particular works of these forerunners as providing insights for their discoveries.

A look at the history shows that Newton and Leibniz had very little of what we would call "calculus results" available to them, and what they did have was fragmentary and unorganized. It was part of their genius that they could see the important facts in the midst of this collection. For example, both men used Pascal's triangle as an important part of their work, although each used it in a different way.

Newton followed the path of an earlier work by Wallis in which interpolation between the rows of Pascal's triangle was used to handle the case when exponents as $\frac{1}{2}, \frac{3}{2}, \frac{5}{2}, \ldots$ were placed between the integer values of

117

$1, 2, 3, \ldots$ in the expansion of $(x + y)^n$. On the other hand, Leibniz used the basic formation rule where each entry is the sum of two entries in the row above it, and formed another triangle where a subtraction formation rule was used for summing certain infinite series.

Newton made his discovery in the period 1665–66, while Cambridge University was closed because of the plague in England. Leibniz made his discovery in the period 1673–76, while on a diplomatic mission in Paris, during which time he made visits to the Royal Society in London and began his first serious study of the mathematical writings of such men as Euclid and Pascal. Newton was very reluctant to publish his findings, and none of his main results were published until after 1700, although he knew these results and used them for over 30 years before publication. Leibniz, on the other hand, published quickly, and although his writings were difficult to comprehend, James and John Bernoulli persisted in their study of them until they understood the new results. Through their writing and teaching, the new results were widely disseminated on the European continent.

Much of this early period of the development of the calculus is marred by a Newton-Leibniz controversy about who discovered the calculus first. Although this disagreement was carried on by their successors, it now seems clear that each man made his discovery independently of the other, using different initial approaches and symbols. Newton used the concept of fluxions and fluents, and his work was carried on by British countrymen such as Brook Taylor and Colin Maclaurin. Leibniz chose the concept of differentials and his approach was successfully expanded, not only by the Bernoullis, but also by the prolific Euler, who seemed to turn everything he touched into a new result in analysis.

Leibniz read about the method by which Pascal had solved a specific problem—namely that of the surface area of a sphere, and saw in the problem a general method in which the surface of a solid of revolution about an axis could be reduced to an equivalent plane figure. This involved the use of what he called the "characteristic triangle." His work with the reciprocals of integers in the harmonic triangle gave values in an easy way for a number of specific series and "implanted in Leibniz's mind a vivid conception that was to play a dominant role in his development of the calculus, namely the notion of an inverse relationship between the operations of taking differences and that of taking sums."[1]

---

[1] *The Historical Development of the Calculus* by C. H. Edwards, Jr., New York: Springer-Verlag, 1979, p. 238.

Leibniz also solved a rectification (i.e. arc length) problem using this characteristic triangle idea to change a rectification problem to a quadrature problem—again by the use of similar triangles. A third application gave the insight for the Fundamental Theorem, or inverse process idea, that $\nu dx = ydy$, where ν is the length of the subnormal. This showed Leibniz how to reduce a quadrature of figures problem to an inverse problem of tangents. Along with his previous work with the harmonic triangle, which showed the "great utility of differences," Leibniz later wrote in a letter to L'Hospital that

> to find quadratures or the sums of the ordinates was the same thing as to find an ordinate of which the difference is proportional to the given ordinate .... to find tangents is nothing else but to find differences, and that to find quadratures is nothing else but to find sums.[2]

―――――――――――――――

[2] *The Historical Introduction of the Calculus* by C. H. Edwards, Jr., New York: Springer-Verlag, 1979, p. 245.

SECTION 19   THE DERIVATIVE AND THE
DIFFERENTIATION RULES

After defining the derivative of a function early in a course of calculus, a few weeks are spent on the task of developing the derivative formulas. These consist of two types—one for general combinations of differentiable functions, such as sums, products, quotients, and composites. The second type consists of derivatives for specific functions, such as the power function x^r, the trigonometric functions and their inverses, the logarithmic and exponential functions, and sometimes the hyperbolic functions. The derivative for a function defined by an integral might also be included.

Typical proofs use the definition of the derivative along with an appropriate algebraic device. For example, the derivative of x^n for $n \in N$ may be found by use of the binomial theorem, while the derivatives for $\sin x$ and $\cos x$ are obtained by using identities, such as those for $\sin(a + b)$ and $\cos(a + b)$, and using geometry to prove that $\lim_{t \to 0} \frac{\sin t}{t} = 1$. The derivatives for the other four trigonometric functions are determined by use of the Quotient Rule, while the derivatives of the inverse trigonometric functions are obtained using the Chain Rule. A common way to find the derivative of $\ln x$ is by use of a function defined by an integral, namely $\int_1^x \frac{1}{t}\, dt$.

One topic omitted from such an approach is a careful existence proof that functions such as $\sin x$ and e^x exist with their commonly assumed properties. A complete study of analysis should include such proofs. In this section, we phrase some of the examples in such a way as to prepare you for these proofs (see Examples 4 and 5).

Definition 1   Let $f : I \to \mathbf{R}$, where I is an interval of real numbers, and let c be an interior point of I. f is *differentiable* at c if $\lim_{x \to c} \frac{f(x) - f(c)}{x - c}$ exists. In this case, the value of this limit is called the *derivative of f at c* and is denoted by the symbol $f'(c)$.

An equivalent form of Definition 1 is to say that f is differentiable at c if $\lim_{h \to 0} \frac{f(c+h) - f(c)}{h}$ exists. The definition can be generalized by assuming that f is defined on a domain set A, where c is a cluster point of A, and that the limit is taken as x approaches c assuming only values in A. In this way, if a function f is defined on the closed interval $[a, b]$, we can define a *one-sided derivative of f at a* by $\lim_{x\to a+}\frac{f(x)-f(a)}{x-a}$, or a *one-sided derivative of f at b* by $\lim_{x\to b-}\frac{f(x)-f(b)}{x-b}$. The derivative f' may then be regarded as a function defined for all x in the interval I. Another symbol for the derivative of f is $\frac{df}{dx}$, which is the differential notation due to Leibniz.

Example 1   Let $f(x) = x^4$. Using Definition 1,

$$f'(c) = \lim_{x\to c}\frac{x^4 - c^4}{x - c} = \lim_{x\to c}(x^3 + x^2c + xc^2 + c^3) = 4c^3$$

Using the alternate form of the definition,

$$f'(c) = \lim_{h\to 0}\frac{(c+h)^4 - c^4}{h}$$

$$= \lim_{h\to 0}\frac{4c^3h + 6c^2h^2 + 4ch^3 + h^4}{h}$$

$$= \lim_{h\to 0}(4c^3 + 6c^2h + 4ch^2 + h^3) = 4c^3$$

Either form of the definition shows that $f'(x)$ exists for all x and is equal to $4x^3$.   □

Example 2   Let the function $f(x) = \sqrt{x}$ be defined on the interval $[0, \infty)$.   Since

$$\lim_{x\to c}\frac{\sqrt{x} - \sqrt{c}}{x - c} = \lim_{x\to c}\frac{1}{\sqrt{x} + \sqrt{c}} = \frac{1}{2\sqrt{c}}$$

for all $c > 0$, f is differentiable for all $x > 0$, and $f'(x) = \frac{1}{2\sqrt{x}}$.   □

Example 3   Let $s(x) = \sin x$ and $c(x) = \cos x$. Then

$$s'(0) = \lim_{x\to 0}\frac{\sin x - \sin 0}{x - 0} = \lim_{x\to 0}\frac{\sin x}{x} = 1$$

This last limit is obtained in a typical calculus course by a geometric argument. For Euler, who defined $\sin x$ by a power series, this fact would be obvious, since if we start with the series expansion for $\sin x$, divide by x, and then take the limit as x approaches 0, we would have

$$\sin x = x - \frac{x^3}{3!} + \frac{x^5}{5!} - + \cdots$$

$$\frac{\sin x}{x} = 1 - \frac{x^2}{3!} + \frac{x^4}{5!} - + \cdots$$

$$\lim_{x \to 0} \frac{\sin x}{x} = \lim_{x \to 0} \left(1 - \frac{x^2}{3!} + \frac{x^4}{5!} - + \cdots \right) = 1$$

In a similar way,

$$c'(0) = \lim_{x \to 0} \frac{\cos x - \cos 0}{x - 0} = \lim_{x \to 0} \frac{\cos x - 1}{x}$$

and this limit is proved to equal 0 by use of the identity $\sin^2 x + \cos^2 x = 1$, and the fact that $\lim_{x \to 0} \frac{\sin x}{x} = 1$. $\square$

Example 4   Let $f : \mathbf{R} \to \mathbf{R}$ be a function that satisfies the following three conditions.

(1) $f(x + y) = f(x) f(y)$ for all $x, y \in \mathbf{R}$

(2) $f(0) = 1$

(3) $f'(0)$ exists

Then $f'(x)$ exists for all $x \in \mathbf{R}$ and equals $f'(0) \cdot f(x)$.

Solution.

$$f'(x) = \lim_{h \to 0} \frac{f(x + h) - f(x)}{h}$$

$$= \lim_{h \to 0} \frac{f(x)f(h) - f(x)}{h}$$

$$= f(x) \lim_{h \to 0} \frac{f(h) - 1}{h}$$

Therefore $f'(x)$ exists, if $\lim_{h \to 0} \frac{f(h) - 1}{h}$ exists. But this limit exists, because by conditions (2) and (3),

$$f'(0) = \lim_{h \to 0} \frac{f(h) - f(0)}{h - 0} = \lim_{h \to 0} \frac{f(h) - 1}{h}   \square$$

Comment. The result in Example 4 will be of interest to us only if there is some useful function that satisfies the three given conditions. You may have observed that $f(x) = b^x$ is a likely candidate, since conditions (1) and (2) are well-known properties for this exponential function. For the present, we appeal to the graph of b^x as intuitive evidence that there is a tangent line to this curve at the point $(0, 1)$, and the slope of this tangent line can be taken as the value for $f'(0)$.

Example 5 Let $s : \mathbf{R} \to \mathbf{R}$ and $c : \mathbf{R} \to \mathbf{R}$ be functions satisfying the following conditions.

(1) $s(x + y) = s(x)c(y) + s(y)c(x)$, for all $x, y \in \mathbf{R}$

(2) $c(x + y) = c(x)c(y) - s(x)s(y)$ for all $x, y \in \mathbf{R}$

(3) $\lim_{x \to 0} \frac{s(x)}{x} = 1$ and $\lim_{x \to 0} \frac{c(x)-1}{x} = 0$

Then $s'(x) = c(x)$ and $c'(x) = -s(x)$ for all $x \in \mathbf{R}$.

Solution.

$$s'(x) = \lim_{h \to 0} \frac{s(x + h) - s(x)}{h}$$

$$= \lim_{h \to 0} \frac{s(x)c(h) + s(h)c(x) - s(x)}{h}$$

$$= s(x) \lim_{h \to 0} \frac{c(h) - 1}{h} + c(x) \lim_{h \to 0} \frac{s(h)}{h}$$

$$= s(x) \cdot 0 + c(x) \cdot 1 = c(x)$$

since the necessary limits are assumed to exist by condition (3). The result involving $c'(x)$ is left as an exercise. Do you know any familiar functions s and c that satisfy the three conditions given in this example? ☐

LEMMA 19.1 If $f : I \to \mathbf{R}$ is differentiable at an interior point c of the interval I, then f is continuous at c.

Proof. The hypothesis implies that $\lim_{x \to c} \frac{f(x) - f(c)}{x - c}$ exists, and equals the real number denoted by $f'(c)$. Then

$$\lim_{x \to c} \left(f(x) - f(c) \right) = \lim_{x \to c} \frac{f(x) - f(c)}{x - c} \cdot (x - c)$$

$$= \lim_{x \to c} \frac{f(x) - f(c)}{x - c} \lim_{x \to c} (x - c)$$

$$= f'(c) \cdot 0 = 0$$

and so f is continuous at c, since $\lim_{x \to c} f(x) = f(c)$.   ∎

Example 6   Since $f(x) = |x|$ is continuous, but not differentiable at $c = 0$, the converse of Lemma 19.1 is not true.   □

THEOREM 19.2 (Limit Theorem for Derivatives)
Let $f : I \to \mathbf{R}$ and $g : I \to \mathbf{R}$ be differentiable at an interior point c of the interval I. Then the following results are true.

(1) $f + g$ is differentiable at c and $(f + g)'(c) = f'(c) + g'(c)$

(2) fg is differentiable at c and $(fg)'(c) = f(c) \cdot g'(c) + f'(c) \cdot g(c)$

(3) $\frac{f}{g}$ is differentiable at c, if $g(c) \neq 0$, and

$$\left(\frac{f}{g}\right)'(c) = \frac{g(c)f'(c) - f(c)g'(c)}{(g(c))^2}$$

We can think of Theorem 19.2 as a limit theorem for derivatives, which makes it the fourth limit theorem in this book. Theorem 4.1 was the limit theorem for sequences, Theorem 10.2 was the limit theorem for functions, and Theorem 15.1 was the limit theorem for continuous functions. You would do well to look over these theorems to observe their similarities. As mentioned earlier, these limit theorems may also be thought of as closure theorems, in that the collections of convergent sequences, continuous functions, and differentiable functions are each closed under the operations of addition, multiplication, and division (when the denominator is non-zero). The proof of Theorem 19.2 uses Definition 1, Lemma 19.1, and Theorem 10.2. We illustrate this by the following proof of part (2).

Proof of (2). fg is differentiable at $x = c$ if $\lim_{x \to c} \frac{f(x)g(x) - f(c)g(c)}{x - c}$ exists. Our work below shows that this limit exists, and that it has the value stated in the theorem. Since g is differentiable at c, we use Lemma 19.1 to assert that g is continuous at c, so that $\lim_{x \to c} g(x) = g(c)$. Then we have

$$\lim_{x \to c} \frac{f(x)g(x) - f(c)g(c)}{x - c} \;=\; \lim_{x \to c} \frac{f(x)g(x) - f(c)g(x) + f(c)g(x) - f(c)g(c)}{x - c}$$

$$= \;\lim_{x \to c} g(x) \lim_{x \to c} \frac{f(x) - f(c)}{x - c}$$

$$+ \lim_{x \to c} f(c) \lim_{x \to c} \frac{g(x) - g(c)}{x - c}$$

$$= \; g(c)\,f'(c) \;+\; f(c)\,g'(c) \quad \blacksquare$$

We come now to one of the most useful derivative formulas—the one used to differentiate composite functions. It has a fairly simple intuitive proof that is usually given in calculus. While this proof is correct for most functions, it is not valid in general. One of the tasks in an analysis course is to show how to adjust these intuitive proofs so that they will be correct for the largest possible collection of functions.

THEOREM 19.3 (The Chain Rule)

If $f : I \to \mathbf{R}$ is differentiable at an interior point c of the interval I, and $g(y)$ is differentiable at $y = f(c)$, then the composite function $(g \circ f)(x)$ is differentiable at c and $(g \circ f)'(c) = g'(f(c)) \cdot f'(c)$.

Proof. The proof approach that is usually seen in a calculus course is to observe that

$$(g \circ f)'(c) \;=\; \lim_{x \to c} \frac{(g \circ f)(x) - (g \circ f)(c)}{x - c}$$

$$= \;\lim_{x \to c} \frac{g(f(x)) - g(f(c))}{f(x) - f(c)} \cdot \frac{f(x) - f(c)}{x - c}$$

$$= \;\lim_{x \to c} \frac{g(f(x)) - g(f(c))}{f(x) - f(c)} \cdot \lim_{x \to c} \frac{f(x) - f(c)}{x - c}$$

$$= \; g'(f(c))\,f'(c)$$

While the second limit is equal to $f'(c)$, the first limit will equal $g'(f(c))$ when the limit is taken as $f(x) \to f(c)$, but in this case, the limit is taken as $x \to c$. In calculus, we argued that as $x \to c$, $f(x) \to f(c)$ by the continuity of f, and so the first limit would equal $g'(f(c))$. This straightforward approach is valid for the majority of functions, but it can fail if $f(x)$ equals $f(c)$ infinitely many times in each neighborhood of c, since a division by 0 is then involved.

To correct this situation, we begin by defining the following function $h(y)$ with a denominator of $y - f(c)$, so that in taking the limit as $y \to f(c)$, we know that $y - f(c) \neq 0$.

$$\text{Let } h(y) \;=\; \begin{cases} \frac{g(y)-g(f(c))}{y-f(c)} & \text{if } y \neq f(c) \\[2mm] g'(f(c)) & \text{if } y = f(c) \end{cases}$$

We have made $h(y)$ continuous at $f(c)$ by this definition, since $h(y)$ at worst has a removable discontinuity at $y = f(c)$. This is so because $\lim_{y \to f(c)} h(y)$ exists, since $g(y)$ is assumed differentiable at $y = f(c)$, and

$$g'(f(c)) \;=\; \lim_{y \to f(c)} \frac{g(y) - g(f(c))}{y - f(c)} \;=\; \lim_{y \to f(c)} h(y)$$

It follows next that

$$g(y) - g(f(c)) \;=\; h(y) \cdot (y - f(c)) \qquad \text{if } y \in D_g$$

$$g(f(x)) - g(f(c)) \;=\; h(f(x)) \Big(f(x) - f(c) \Big) \qquad \text{if } x \in D_{g \circ f}$$

$$\frac{g(f(x)) - g(f(c))}{x - c} \;=\; h(f(x)) \frac{f(x) - f(c)}{x - c} \qquad \text{if } x \neq c$$

Therefore

$$\lim_{x \to c} \frac{g(f(x)) - g(f(c))}{x - c} \;=\; \lim_{x \to c} h(f(x)) \cdot \lim_{x \to c} \frac{f(x) - f(c)}{x - c}$$

and by rewriting these three limits,

$$(g \circ f)'(c) \;=\; g'(f(c)) \cdot f'(c) \qquad \blacksquare$$

Example 7   By using 2 or more of the derivative results proved in this section, we can obtain several additional derivative formulas such as the following three, as well as the results listed in Exercise 9.

(a) Using $\frac{d}{dx}(\sin x) = \cos x$, and the Quotient Rule, we know that

$$\frac{d}{dx}(\csc x) = \frac{d}{dx}\left(\frac{1}{\sin x}\right) = -\frac{\cos x}{\sin^2 x} = -\csc x \cot x$$

(b) Using $\frac{d}{dy}(\sin y) = \cos y$, and the Chain Rule, and assuming that $y = \arcsin x$ is differentiable, we can show that

$$\frac{d}{dx}(\arcsin x) = \frac{1}{\sqrt{1 - x^2}}$$

For if $y = \arcsin x$, $x = \sin y$, and so $1 = \cos y \frac{dy}{dx}$ or $\frac{dy}{dx} = \frac{1}{\cos y}$.
Since $\cos y = \sqrt{1 - \sin^2 y} = \sqrt{1 - x^2}$, and since the slope of the tangent line to the graph of arcsin x is always positive, the derivation is complete.

(c) The power rule can be extended to negative integer exponents since

$$\frac{d}{dx}(x^{-n}) = \frac{d}{dx}\left(\frac{1}{x^n}\right) = \frac{(x^n)0 - 1(nx^{n-1})}{x^{2n}} = -nx^{-n-1} \qquad \square$$

Example 8   If possible, find an expression for the nth derivative $f^{(n)}(a)$ of the following functions. This information will be helpful when we discuss Taylor's Theorem, and the representation of functions by infinite Taylor series expansions.

(a) Let $f(x) = \sin x$ and $a = 0$. Then

$$
\begin{aligned}
f'(x) &= \cos x, &\text{so}\quad f'(0) &= 1\\
f''(x) &= -\sin x, &\text{so}\quad f''(0) &= 0\\
f^{(3)}(x) &= -\cos x, &\text{so}\quad f^{(3)}(0) &= -1\\
f^{(4)}(x) &= \sin x, &\text{so}\quad f^{(4)}(0) &= 0
\end{aligned}
$$

and the pattern of $1, 0, -1, 0$ repeats in blocks of four. So we can express $f^{(n)}(0)$ as follows. Let $f^{(2n-1)}(0) = (-1)^{n+1}$ and $f^{(2n)}(0) = 0$.

(b) Let $f(x) = \ln x$ and $a = 1$. Then we have

$$f'(x) = \frac{1}{x}$$

$$f^{(2)}(x) = -\frac{1}{x^2}$$

$$f^{(3)}(x) = \frac{2}{x^3}$$

$$f^{(4)}(x) = -\frac{6}{x^4}$$

$$f^{(5)}(x) = \frac{24}{x^5}$$

From the pattern developing in these derivatives, we conjecture that

$$f^{(n)}(x) = (-1)^{n-1}\frac{(n-1)!}{x^n}$$

so that

$$f^{(n)}(1) = (-1)^{n-1}(n-1)!$$

(c) Let $f(x) = \tan x$ and $a = 0$. Then

$$f'(x) = \sec^2 x$$
$$f''(x) = 2\sec^2 x \tan x$$
$$f'''(x) = 2\sec^4 x + 4\tan^2 x \sec^2 x$$

As more derivatives are taken, the expressions become increasingly complicated and no general pattern emerges. □

EXERCISES 19

1. Use the definition of the derivative to find f' for the following functions.
   (a) $f(x) = x^5$
   (b) $f(x) = \frac{1}{x^2}$
   (c) $f(x) = |x|$
   (d) $f(x) = \sqrt[3]{x}$
   (e) $f(x) = \frac{1}{x-1}$
   (f) $f(x) = \frac{1}{x^2+4}$
   (g) $f(x) = \sqrt{2-x}$

2. Derive the formula $c'(x) = c'(0)c(x) - s'(0)s(x)$ as given in Example 5.

3. For $f(x) = \log_b x$, use the derivative definition and logarithmic identities to prove that $f'(x) = \frac{1}{x}f'(1)$.

4. Prove Result (1) of Theorem 19.2.

5. Prove Result (3) of Theorem 19.2.

6. Prove that $f(x) = \begin{cases} x\sin\frac{1}{x} & \text{if } x \neq 0 \\ 0 & \text{if } x = 0 \end{cases}$ is continuous but not differentiable at 0.

7. By setting $x = \frac{1}{n}$, find a sequence that converges to the same value as $\lim_{x\to 0+}\frac{b^x-1}{x}$, where $b > 0$.

8. If $f(x) = x^n$ for $n \in \mathbf{N}$, prove that $f'(x) = nx^{n-1}$. You will need an identity for $(x+h)^n$ or $x^n - c^n$, depending on which form of the definition you use.

9. Use 2 or more of the derivative results proved in this section to derive the following formulas. See Example 7. Assume that the derivatives exist in parts (a), (c), and (f).

   (a) $\frac{d}{dx}(x^r) = rx^{r-1}$ where r is the rational number $r = \frac{p}{q}$, and p and q are positive integers.

   (b) $\frac{d}{dx}(\tan x) = \sec^2 x$

   (c) $\frac{d}{dx}\arctan x = \frac{1}{1+x^2}$

   (d) $\frac{d}{dx}\log_b x = \frac{1}{x}\cdot\lim_{h\to 0}\frac{h}{b^h-1}$ Hint: See Example 4.

   (e) $\frac{d}{dx}(\sec x) = \sec x \tan x$

   (f) $\frac{d}{dx}(\operatorname{arcsec} x) = \frac{1}{|x|\sqrt{x^2-1}}$

10. Calculate the following derivatives by use of the results obtained in this section.

    (a) $\frac{d}{dx}(x^3\cos\frac{1}{x})$

    (b) $\frac{d}{dx}(e^{-\frac{1}{x^2}})$

    (c) $\frac{d}{dx}(\arctan\sqrt{x})$

    (d) $\frac{d}{dx}\log_b\sqrt{x^2+1}$

11. (a) Follow the hint in Example 3 to show that $\lim_{x\to 0}\frac{\cos x-1}{x} = 0$.

    (b) Show also how Euler would have obtained this limit value, starting with the power series $\cos x = 1 - \frac{x^2}{2!} + \frac{x^4}{4!} - + \cdots$

12. In Example 5, show that condition (3) can be replaced by the following condition (3'): $s(0) = 0$, $s'(0) = 1$, $c(0) = 1$, and $c'(0) = 0$.

13. If possible, find an expression for the nth derivative $f^{(n)}(a)$, for each of the following functions at the given value of a.

(a) $f(x) = \sin x$ at $a = \pi$

(b) $f(x) = \cos x$ at $a = 0$

(c) $f(x) = e^x$ at $a = 0$

(d) $f(x) = \sqrt{x}$ at $a = 1$

(e) $f(x) = \arctan x$ at $a = 0$

14. The function f is called *periodic with period p* if $f(x + p) = f(x)$ for all $x \in D_f$ so that $x + p \in D_f$ also. For example, $\sin x$ and $\cos x$ are periodic with period 2π, while $\tan x$ is periodic with period π. Prove that if f is differentiable and periodic with period p, then f' is also periodic with period p.

15. Verify that $f(x) = x - [x]$ is periodic, using the definition in Exercise 14.

SECTION 20   THE MEAN VALUE THEOREM

In Chapter 4, we saw that the Bolzano-Weierstrass Theorem could be used to prove a number of results about continuous functions, in particular, Theorems 16.1, 16.2, and 17.2. In this section, we see that Rolle's Theorem plays a very similar role, in that we use it to prove a number of other results, many of which are well-known and useful calculus results. Some of these are shown in the chart below.

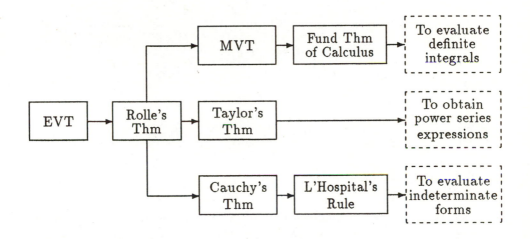

Figure 20.1

Definition 1   The function $f : I \to \mathbf{R}$ has a *relative minimum* at $c \in I$ if there is a neighborhood U of c such that $f(x) \geq f(c)$ for all $x \in U \cap I$. Similarly, f has a *relative maximum* at $c \in I$ if there is a neighborhood U of c such that $f(x) \leq f(c)$ for all $x \in U \cap I$.

LEMMA 20.1   Let $f : I \to \mathbf{R}$ have a relative maximum or minimum at $x = c$, and assume that $f'(c)$ exists. Then $f'(c) = 0$.

Proof.   Assume that f has a relative minimum at the point $c \in I$. It follows that for all x in some interval of c,

$$\frac{f(x) - f(c)}{x - c} \geq 0 \qquad \text{for } x > c$$

$$\frac{f(x) - f(c)}{x - c} \leq 0 \qquad \text{for } x < c$$

Therefore we have

$$\lim_{x \to c^+} \frac{f(x) - f(c)}{x - c} \geq 0$$

$$\lim_{x \to c^-} \frac{f(x) - f(c)}{x - c} \leq 0$$

Since $f'(c)$ exists, it follows that

$$f'(c) = \lim_{x \to c^+} \frac{f(x) - f(c)}{x - c} = \lim_{x \to c^-} \frac{f(x) - f(c)}{x - c}$$

and so $f'(c) = 0$. ∎

A similar proof can be given if f has a relative maximum at the point c. See the exercises. It is possible that f could have a relative minimum or maximum at c without having $f'(c) = 0$. $f(x) = |x|$ at $c = 0$ is an example of this—notice that this does not contradict Lemma 20.1.

THEOREM 20.2 (Rolle's Theorem)

If f is continuous on $[a, b]$, differentiable on (a, b), and $f(a) = f(b) = 0$, then there is some value $c \in (a, b)$ so that $f'(c) = 0$.

Proof. We first consider the trivial case where $f(x) = 0$ for all $x \in [a, b]$. The theorem is true in this case because $f'(c) = 0$ for all $c \in (a, b)$.

Otherwise, there must be at least one $x \in (a, b)$ where $f(x) \neq 0$. If we assume that $f(x) > 0$ for some $x \in (a, b)$, then $M = \sup\{f(x) : x \in [a, b]\}$ is a positive finite number, and there must be a value $c \in (a, b)$, so that $f(c) = M$. Why? It then follows from Lemma 20.1 that $f'(c) = 0$.

If $f(x) < 0$ for some $x \in (a, b)$, then a similar argument can be used involving $m = \inf\{f(x) : x \in [a, b]\}$. ∎

THEOREM 20.3 (Mean Value Theorem)
If f is continuous on $[a, b]$ and differentiable on (a, b), then there is at least one value $c \in (a, b)$ such that

$$f'(c) = \frac{f(b) - f(a)}{b - a}$$

Comment.   It is fortunate that the Mean Value Theorem has a simple geometric interpretation—namely, if the graph of f has no breaks in it and no sharp points, then at least one tangent line to f must be parallel to the line joining the endpoints of the graph (see Figure 20.2). This line is sometimes called the secant line. The proof itself is not intuitive however, since we seem to apply Rolle's Theorem to a strange function in order to come up with the desired result. This "strangeness" disappears when we consider the function $L(x)$ defined in Exercise 13.

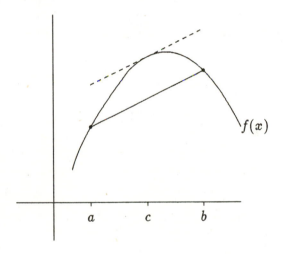

$f(x)$

a c b

Figure 20.2

Proof.   We define the function

$$F(x) = f(x) - f(a) - \frac{f(b) - f(a)}{b - a}(x - a)$$

and assert that F satisfies the hypotheses of Rolle's Theorem. You should verify that F is continuous on $[a, b]$, differentiable on (a, b), and such that $F(a) = F(b) = 0$. Since $F'(x) = f'(x) - \frac{f(b)-f(a)}{b-a}$, the conclusion that $F'(c) = 0$ for some $c \in (a,b)$ is equivalent to the statement that $f'(c) = \frac{f(b)-f(a)}{b-a}$, and the proof is complete. ∎

We next consider some results that can be proved as corollaries of the Mean Value Theorem.

COROLLARY 20.4 If $f' > 0$ on an interval I , then f is an increasing function on I. The interval I can be open or closed, finite or infinite.

Proof. Let $x < u$ on the interval I. Since $f' > 0$ on $[x, u]$, f is certainly continuous and differentiable on $[x, u]$, so by the Mean Value Theorem, $\frac{f(u)-f(x)}{u-x} = f'(c) > 0$ for some $c \in (x, u)$. Then $f(u) - f(x) > 0$ or $f(u) > f(x)$, so f is increasing on I. ∎

Example 1 The function $f(x) = \sin x$ has the property that $f'(x) > 0$ on the set $S = (0, \frac{\pi}{2}) \cup (\frac{3\pi}{2}, 2\pi)$, but f is not increasing on this set. Why is this not a contradiction of Corollary 20.4? □

COROLLARY 20.5 If $f' < 0$ on an interval I, then f is decreasing on the interval I.

Proof. This proof is left as an exercise.

COROLLARY 20.6 If f has a bounded derivative on an interval I, then f is uniformly continuous on I.

Proof. Let $|f'(x)| \leq B$ for all $x \in I$. Then $\delta = \frac{\epsilon}{B}$ is a delta of uniform continuity for f on I. For if $x, u \in I$ and $|x - u| < \delta$, then by the Mean Value Theorem, there is some $c \in (x, u)$ so that

$$\frac{f(x) - f(u)}{x - u} = f'(c)$$

or $|f(x) - f(u)| = |f'(c)| |x - u| \leq B \delta = \epsilon$ ∎

This last result is an interesting extension of Lemma 19.1, and it provides a way to check for uniform continuity on sets other than closed intervals. You should notice that this corollary provides a partial answer to Conjecture 4 as listed at the end of Section 17.

COROLLARY 20.7 If $f' = g'$ on $[a, b]$, then there is some constant C so that $f(x) = g(x) + C$, for all $x \in [a, b]$.

Proof. This is left as an exercise.

Example 2 The sequence $\{s_n\}$, where $s_n = \frac{\ln n}{n}$, is decreasing for all $n \geq 3$. This follows from Corollary 20.5, because the function $f(x) = \frac{\ln x}{x}$ is decreasing for $x > e$, since $f'(x) = \frac{1 - \ln x}{x^2} < 0$ when $1 < \ln x$, or when $x > e$. This information will be helpful when we apply the Alternating Series Test to $\sum (-1)^{n-1} \frac{\ln n}{n}$ in Chapter 8. $\square$

Example 3 $f(x) = \frac{1}{x^2}$ is uniformly continuous on the set $S = [c, \infty)$ for $c > 0$, because of Corollary 20.6, and the fact that for all $x \in S$,

$$|f'(x)| = \left| \frac{-2}{x^3} \right| \leq \frac{2}{c^3} \quad \square$$

Example 4 The function $f(x) = x^2$ is uniformly continuous on the set $S = [0, b]$, since $|f'(x)| = |2x| \leq 2b$ on the set S. It is also uniformly continuous on S because of Theorem 17.2. $\square$

Example 5 The function $f(x) = \sqrt{x}$ is uniformly continuous on $[0, 1]$ by Theorem 17.2, because f is continuous on a closed interval. Further, $f(x) = \sqrt{x}$ is uniformly continuous on $[1, \infty)$ by Corollary 20.6, and the fact that $|f'(x)| = |\frac{1}{2\sqrt{x}}| \leq \frac{1}{2}$ on $[1, \infty)$. Does this imply that $f(x) = \sqrt{x}$ is uniformly continuous on $[0, \infty) = [0, 1] \cup [1, \infty)$? More generally, is it true that if f is uniformly continuous on $[a, b]$ and $[b, c]$, then f must be uniformly continuous on $[a, c]$? $\square$

Example 6 Use the Mean Value Theorem to find upper and lower bounds for $\sqrt{67}$.

Solution. By applying the Mean Value Theorem to $f(x) = \sqrt{x}$ on the interval $[64, 67]$, we have $\frac{\sqrt{67} - \sqrt{64}}{67 - 64} = \frac{1}{2\sqrt{c}}$ for some value c for which $64 < c < 67 < 81$. This leads to the following string of inequalities.

$$\frac{1}{18} < \frac{1}{2\sqrt{c}} < \frac{1}{16}$$

$$\frac{1}{18} < \frac{\sqrt{67} - 8}{3} < \frac{1}{16}$$

$$8\frac{1}{6} < \sqrt{67} < 8\frac{3}{16}$$

$$8.16666 < \sqrt{67} < 8.1875 \quad \square$$

Example 7 Illustrate the Mean Value Theorem for $f(x) = x^3 + 4x^2 - 2$ on $[-1, 2]$.

Solution. We need to find some value $c \in (-1, 2)$, so that

$$\frac{f(2) - f(-1)}{2 - (-1)} = f'(c)$$

This equation becomes $3c^2 + 8c = 7$, for which the solutions are $c = \frac{-4 + \sqrt{37}}{3} \approx .69$ and $c = \frac{-4 - \sqrt{37}}{3} \approx -3.36$. While the tangent lines to the curve at both points are parallel to the secant line, only the point $\frac{-4 + \sqrt{37}}{3}$ is in the interval $(-1, 2)$. See Figure 20.3. $\square$

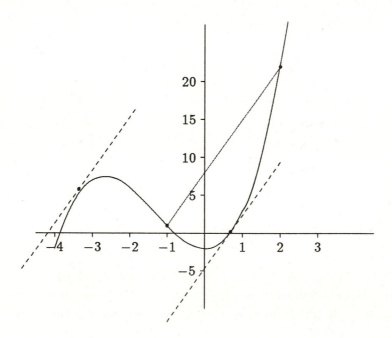

Figure 20.3

Example 8   Show that if f and f' are continuous on $[a, b]$ and f'' exists on (a, b), then for each $x \in [a, b]$, there is a value of $c \in (a, x)$, so that

$$f(x) = f(a) + f'(a)(x - a) + \frac{f''(c)}{2!}(x - a)^2$$

This is the case $n = 1$ of Taylor's Theorem in the next section.

Solution.   Define $F(t) = f(x) - f(t) - (x - t)f'(t)$, where x is considered to be fixed, and show that $F'(t) = -(x - t)f''(t)$. Then define

$$G(t) = F(t) - \left(\frac{x - t}{x - a}\right)^2 F(a)$$

and show that $G(a) = 0$ and $G(x) = 0$. By applying Rolle's Theorem to $G(x)$ on $[a, x]$, it follows that $G'(c) = 0$ for some value c between a and x. Since

$$G'(t) = F'(t) + \frac{2(x - t)}{(x - a)^2} F(a)$$

you can show in a few algebraic steps that $G'(c) = 0$ is equivalent to the desired conclusion.   □

EXERCISES 20

1. Illustrate the Mean Value Theorem for the following functions.
   (a) $f(x) = x^2 + 3x + 4$ on $[1, 4]$
   (b) $f(x) = \ln x$ on $[1, e]$
   (c) $f(x) = \arcsin x$ on $[\frac{1}{2}, 1]$
   (d) $f(x) = \sqrt{x}$ on $[a, b]$

2. For which x will the following functions be increasing? be decreasing?
   (a) $f(x) = x^3 - 3x^2 - 9x$
   (b) $f(x) = \frac{x^2 + 2}{2x - 1}$
   (c) $f(x) = \frac{\ln x}{x^2}$

3. Use Corollary 20.6 and/or Theorem 17.2 to show that the following functions are uniformly continuous on the given set.
   (a) $f(x) = \frac{1}{\sqrt{x}}$ on $[c, \infty)$, where $c > 0$
   (b) $f(x) = \arctan x$ on $(-\infty, \infty)$
   (c) $f(x) = x^{\frac{3}{2}}$ on $[0, c]$

4. On which sets will these functions be uniformly continuous?

   (a) $f(x) = \frac{1}{x+1}$

   (b) $f(x) = \frac{1}{x^2+1}$

   (c) $f(x) = \sin\frac{1}{x}$

   (d) $f(x) = \tan x$

5. Use the Mean Value Theorem to find upper and lower bounds for the following values.

   (a) $\sqrt{47}$

   (b) $\sqrt[3]{31}$

   (c) $\ln .5$

   (d) $\sin 1.5$

6. Give an example of a function which satisfies each of the following conditions.

   (a) A function f that is continuous on $[a, b]$ and differentiable on (a, b), but not differentiable on $[a, b]$.

   (b) A function f that is continuous and differentiable on (a, b), but not on $[a, b]$, and such that the conclusion of the Mean Value Theorem does apply.

   (c) A function f that is continuous and differentiable on (a, b), but not on $[a, b]$, and such that the conclusion of the Mean Value Theorem does not apply.

   (d) A function f that is continuous on $[a, b]$, not differentiable on (a, b), and is such that the conclusion of the Mean Value Theorem does not apply.

7. Find an approximation for $\sqrt{67}$ by these two methods from calculus and compare them with the approximation obtained in Example 6 by the use of the Mean Value Theorem.

   (a) Differentials

   (b) Newton's Method

8. Prove Lemma 20.1 for the case where f has a relative maximum at $x = c$.

9. Prove Rolle's Theorem for the case where $f(x) < 0$, for some $x \in (a, b)$.

10. Prove Corollary 20.5.

11. Prove Corollary 20.7. Hint: Let $h(x) = f(x) - g(x)$, and show that h is a constant function.

12. Answer the question posed at the end of Example 5.

13. Let the function f satisfy the conditions of the Mean Value Theorem.

    (a) Find the equation of the straight line function $L(x)$ joining the points $(a, f(a))$ and $(b, f(b))$.

    (b) Show that $F(x) = f(x) - L(x)$ is the function used in the proof of the Mean Value Theorem.

14. Use the Intermediate Value Theorem to prove that the polynomial function $f(x) = x^5 + 2x^3 + x - 3$ has a real zero. Then use Rolle's Theorem to show that f has only one real zero.

15. Provide details for the steps in the solution provided for Example 8.

16. Write a one paragraph essay describing the process by which we can prove the Mean Value Theorem, given the Extreme Value Theorem.

SECTION 21   TAYLOR'S THEOREM WITH REMAINDER

Students usually agree that the unit on infinite series is one of the most difficult in calculus. The fact that it comes at the very end of the material on functions of one variable may add to the evidence for such a conviction (i.e. the hardest topic is saved for the last). However, in the seventeenth and eighteenth centuries, this was not the case. During this time, the expression of functions in the form of infinite polynomials occurred at the very beginning of a study of calculus, and derivatives and integrals of functions were calculated in terms of these series expressions. After all, what could be easier than working with polynomials! There was none of our modern concern with uniform convergence, or intervals of convergence, or similar theoretical matters. In fact, an early preoccupation with the correct theory of convergence might well have stifled the excitement and creativity that quickly led to the tools of calculus that are so useful in answering questions about our physical world.

When the time comes to provide a proper theoretical basis for these infinite polynomial expressions known as Taylor or Maclaurin series, such as

$$e^x = 1 + x + \frac{x^2}{2!} + \frac{x^3}{3!} + \cdots + \frac{x^n}{n!} + \cdots$$

then Taylor's Theorem with Remainder plays a crucial role. In this section, we give a proof for this important result. Historically, this topic arose as an interpolation question through the attempt to construct a table of logarithm values. Because no computers or hand-held calculators were available in those early years, it was desirable to minimize the number of values to be calculated, with interpolation and log identities used to determine in-between values. For example, if the goal was to construct a table for $\log x$ from $x = 1$ to $x = 100$ in increments of 0.1, it would be necessary only to calculate $\log x$ for the 25 primes between 1 and 100. Then logs of composites could be found by identities such as

$$\log 54 = \log(2 \cdot 3^3) = \log 2 + 3 \log 3 \quad \text{and} \quad \log 55 = \log 5 + \log 11$$

Finally, the values for $\log 54.1$, $\log 54.2$, $\ldots$, $\log 54.9$ could be determined by interpolation between $\log 54$ and $\log 55$. However, the simple process of straight line interpolation is inaccurate here, so refinements were needed.

Around 1620, Henry Briggs showed how to use the first and second difference approach to get values accurate to 5 decimal places. Such techniques
are studied in a numerical analysis course.

When this approximation method is generalized, as it was by Newton
in the *Principia* and by James Gregory in a letter of 1670, it leads to a
form of what we today call Taylor's Theorem, so named because Brook
Taylor first published the Taylor series formula in 1715. Unfortunately for
Taylor, John Bernoulli published a result twenty years earlier in the *Acta
Eruditorum* of 1694, which so closely resembled the result of Taylor that
Bernoulli accused him of plagiarism.

As Newton stated in his *Principia,* the purpose of this approximation
work is "to find a curved line of the parabolic kind (by this he meant
polynomial) which shall pass through any given number of points." The
approach used by Newton and Gregory was to find the coefficients of the
polynomial

$$p(x) = a_0 + a_1x + a_2x^2 + \cdots + a_nx^n$$

that agrees with a given function f at $n+1$ equally spaced points along
the x-axis. Then interpolation is used to approximate the value at any
intermediate point.

Another goal of this work was to approximate areas under curves by
finding the area under an approximating polynomial. The familiar trape-
zoidal rule comes from the approximation

$$\int_{x_0}^{x_1} f(x)\,dx \approx \frac{\Delta x}{2}(y_0 + y_1)$$

which is obtained when we use a first degree polynomial that is determined
by the two points x_0 and x_1. When second degree polynomials determined
by the three equally spaced points x_0, x_1, and x_2 are used, we obtain
Simpson's parabolic rule from the approximation

$$\int_{x_0}^{x_2} f(x)\,dx \approx \frac{\Delta x}{3}(y_0 + 4y_1 + y_2)$$

A further result that is rarely studied in calculus is the "Newton-Cotes
three-eighths rule" that

$$\int_{x_0}^{x_3} f(x)\,dx \approx \frac{3\Delta x}{8}(y_0 + 3y_1 + 3y_2 + y_3)$$

which is obtained by using a polynomial of third degree that is determined by the four equally spaced points x_0, x_1, x_2, and x_3. The polynomial approximation method developed by Newton and his successors generalizes these results for the case when we use a polynomial of degree n, where n is any positive integer.

The early derivations for the coefficients of the Taylor series approximations represent a variety of methods, and are difficult to understand because of complicated notation. These methods did not include the remainder term that we now use to measure the error, or the use of the Mean Value Theorem to make the proof shorter and simpler. However, a modern proof often lacks motivation and provides only a "surgically-clean" proof.

THEOREM 21.1 (Taylor's Theorem with Remainder)
Let $f(t)$ be a function whose first n derivatives are continuous on $[a, b]$, and is such that $f^{(n+1)}(t)$ exists on (a, b). Define

$$P_n(t) = f(a) + f'(a)(t - a) + \cdots + \frac{f^{(n)}(a)}{n!}(t - a)^n$$

Then for each value of $x \in (a, b)$, there is some value c between a and x so that

$$f(x) = P_n(x) + R_n(c, x)$$

where
$$R_n(c, x) = \frac{f^{(n+1)}(c)}{(n + 1)!}(x - a)^{n+1}$$

Comment. $R_n(c, x)$ is called the *remainder term*. $P_n(x)$ is the *Taylor polynomial of degree* n, and it approximates the value of $f(x)$ with an error given by R_n. This theorem is really an infinite collection of results, all of which can be established by a single proof. It may help to understand the proof of this important result better if a few special cases are considered first. If $n = 0$, Taylor's Theorem becomes the Mean Value Theorem, since

$$f(x) = P_0(x) + R_0(x) = f(a) + f'(c) \cdot (x - a)$$

reduces to
$$\frac{f(x) - f(a)}{(x - a)} = f'(c)$$

The case where $n = 1$ was done as Example 20.8, while the case where $n = 2$ is left as an exercise. Working through these special cases will help you to better understand the general proof that follows, because the method of proof is identical in all cases. However, no geometric interpretation or motivation can be given for this proof.

Proof. Considering x to be fixed, we define the function

$$F(t) \ = \ f(x) - f(t) - (x - t)f'(t) - \frac{(x - t)^2}{2!}f''(t) - \cdots - \frac{(x - t)^n}{n!}f^{(n)}(t)$$

In calculating $F'(t)$, we obtain 2n+1 terms, and they all cancel out but one, leaving us with $F'(t) = -\frac{(x-t)^n}{n!}f^{(n+1)}(t)$. We next define

$$G(t) \ = \ F(t) - \left(\frac{x - t}{x - a}\right)^{n+1}F(a)$$

and show that $G(a) = G(x) = 0$. This permits us to apply Rolle's Theorem to $G(t)$ on $[a, x]$, from which we have,

$$G'(c) \ = \ F'(c) + \frac{(n+1)}{x - a}\left(\frac{x - c}{x - a}\right)^n F(a) \ = \ 0$$

for some value $c \in (a, x)$. By substituting in the values for $F(a)$ and $F'(c)$, you may obtain the desired result in a few steps, and are encouraged to work through the details as an exercise. ∎

COROLLARY 21.2   If $f^{(n)}(x)$ exists for all n, and for all $x \in [a, b]$, and if $\lim_{n \to \infty} R_n(x_0) = 0$ for $x_0 \in [a, b]$, then the sequence $\{P_n(x_0)\}$ converges to $f(x_0)$, and we call

$$\lim_{n \to \infty} P_n(x_0) \ = \ \sum_{k=0}^{\infty} \frac{f^{(k)}(a)}{k!}(x_0 - a)^k$$

the Taylor series expression for f at x_o in powers of $(x_o - a)$.

Example 1   Use Taylor's Theorem with $n = 1$ to find upper and lower bounds for $\sqrt{67}$.

Solution.   We set $f(t) = \sqrt{t}$, $a = 64$, $x = 67$, and use

$$f'(t) \ = \ \frac{1}{2\sqrt{t}} \quad \text{and} \quad f''(t) \ = \ \frac{-1}{4t^{3/2}}$$

to obtain the equation

$$\sqrt{67} = \sqrt{64} + \frac{1}{2\sqrt{64}}(67-64) - \frac{1}{8c^{3/2}}(67-64)^2$$

$$= 8 + \frac{3}{16} - \frac{9}{8c^{3/2}}$$

for some value c so that $64 < c < 67 < 81$. This can be used to obtain the following set of inequalities.

$$\frac{1}{8 \cdot 9^2} < \frac{9}{8c^{3/2}} < \frac{9}{8^4}$$

$$-\frac{9}{8^4} < \sqrt{67} - 8\frac{3}{16} < -\frac{1}{8 \cdot 9^2}$$

$$8\frac{3}{16} - \frac{9}{4096} < \sqrt{67} < 8\frac{3}{16} - \frac{1}{648}$$

$$8.1853 < \sqrt{67} < 8.18596$$

This gives an approximation for $\sqrt{67}$ that is accurate to 3 decimal places, while the case when $n = 0$ gives only 1-decimal-place accuracy (see Example 20.6). However, an algebraically simpler approach is the Maclaurin series for $\sqrt{x+1}$ (see Example 6).   □

Example 2   Apply Taylor's Theorem to $f(x) = \ln x$, using $a = 1$ and $n = 3$ (see Example 19.8b). Then use this result to find a value for $\ln 3$.

Solution.   After calculating the first four derivatives of $\ln x$, we can write

$$\ln x = (x-1) - \frac{1}{2}(x-1)^2 + \frac{1}{3}(x-1)^3 - \frac{1}{4c^4}(x-1)^4$$

for some value c between 1 and x. By setting $x = 3$, we have $\ln 3 = \frac{8}{3} - \frac{4}{c^4}$ for some value $c \in (1,3)$. Since $\frac{4}{81} < \frac{4}{c^4} < 4$ when $1 < c < 3$, we can only say that $\ln 3 \approx \frac{8}{3}$ with an error less than 4, which is not a very helpful statement. It will not make things better to choose a larger value for n, since the remainder term of $\frac{2^n}{nc^n}$ has no upper bound for $1 < c < 3$. The problem here is that $x = 3$ is not within the interval of convergence for the series $\sum(-1)^{n-1}\frac{(x-1)^n}{n}$. We will see that there are ways around this difficulty, such as using $x = \frac{1}{3}$, since $\ln\frac{1}{3} = -\ln 3$.   □

Example 3   Find $P_n(x)$ and $R_n(x)$ when Taylor's Theorem is applied to $f(x) = \cos x$ using $a = 0$.

Solution.   In order to apply Taylor's Theorem, we need to find an expression for $f^{(n)}(a)$. Because the derivatives of $\cos x$ repeat in blocks of 4, we know that the sequence $\{f^{(n)}(0)\}$ is $\{1, 0, -1, 0, 1, 0, -1, 0, \ldots\}$, starting with $n = 0$. Therefore,

$$P_{2n}(x) \;=\; 1 - \frac{x^2}{2!} + \frac{x^4}{4!} - \cdots (-1)^n \frac{x^{2n}}{(2n)!}$$

$$R_{2n}(x) \;=\; \frac{f^{(2n+1)}(c)}{(2n+1)!}(x - 0)^{2n+1} \;=\; \frac{(\pm \sin c)\, x^{2n+1}}{(2n+1)!}$$

for some value of c between 0 and x.   □

Example 4   Find the values of x so that $\lim R_n(x) = 0$ in Example 3.

Solution.   For any value of c, $|\pm \sin c| \le 1$, and so by Exercise 4.20,

$$\lim_{n \to \infty} |R_n(x)| \;\le\; \lim_{n \to \infty} \frac{|x^{2n+1}|}{(2n+1)!} \;=\; 0$$

for any real number x. This fact constitutes a proof that the series

$$1 - \frac{x^2}{2!} + \frac{x^4}{4!} - \cdots + (-1)^n \frac{x^{2n}}{(2n)!} + \cdots$$

converges to $\cos x$, for all real numbers x.   □

Example 5   Find $P_n(x)$ and $R_n(x)$ when Taylor's Theorem is applied to $f(x) = \sqrt{x+1}$ with $a = 0$.

Solution.   Though it is a little complicated, there is a discernible pattern for the sequence $\{f^{(n)}(0)\}$. Since

$$f'(x) \;=\; \frac{1}{2}(x+1)^{-\frac{1}{2}}$$

$$f''(x) \;=\; \frac{1}{2} \cdot \frac{-1}{2}(x+1)^{-\frac{3}{2}}$$

$$f'''(x) \;=\; \frac{1}{2} \cdot \frac{-1}{2} \cdot \frac{-3}{2}(x+1)^{-\frac{5}{2}}$$

we conjecture that

$$f^{(n)}(x) = (-1)^{n-1}\frac{1\cdot 3\cdot 5\cdots(2n-3)}{2^n}(x+1)^{-\frac{2n-1}{2}}$$

$$f^{(n)}(0) = (-1)^{n-1}\frac{1\cdot 3\cdot 5\cdots(2n-3)}{2^n}$$

Therefore the Taylor polynomial is

$$P_n(x) = \sum_{k=0}^{n}\frac{f^{(k)}(0)}{k!}(x-0)^k$$

$$= 1+\frac{1}{2}x-\frac{1}{8}x^2+\frac{1}{16}x^3-\frac{5}{128}x^4+\cdots$$

$$\cdots+(-1)^{n-1}\frac{1\cdot 3\cdots(2n-3)}{2^n\cdot n!}x^n$$

and the Remainder term is

$$R_n(x) = \frac{f^{(n+1)}(c)}{(n+1)!}(x-0)^{n+1}$$

$$= (-1)^n\frac{1\cdot 3\cdot 5\cdots(2n-1)x^{n+1}}{2^n(n+1)!}(c+1)^{-\frac{2n+1}{2}}$$

This remainder term is clearly more complicated than the remainder term for $\cos x$ in Example 3, making it difficult to show that $\lim_{n\to\infty} R_n(x) = 0$ for $|x| < 1$.   □

Example 6   Use the result in Example 5 to approximate $\sqrt{67}$.

Solution.   From Example 5, we have the result that

$$\sqrt{x+1} = 1+\frac{1}{2}x-\frac{1}{8}x^2+\frac{1}{16}x^3-\cdots+(-1)^{n-1}\frac{1\cdot 3\cdots(2n-3)}{2^n\cdot n!}x^n$$

$$+(-1)^n\frac{1\cdot 3\cdots(2n-1)}{2^n(n+1)!}(c+1)^{-\frac{2n+1}{2}}x^{n+1}$$

for some value c between 0 and x. It turns out that the choice of $x = 66$ is disastrous here because

$$\sqrt{67} = 1 + \frac{66}{2} - \frac{66^2}{8} + \frac{66^3}{16} + R_3(66)$$

$$= 1 + 33 - 544.5 + 17968.5 + R_3(66)$$

$$= 17458 + R_3(66)$$

indicates that the error term is very large and will only become worse for larger values of n. In Chapter 9, we see that the remainder term will approach zero, and the polynomial series will converge if x is chosen so that $|x| < 1$. So if we set $x = \frac{3}{64}$, we obtain a convergent series expression for $\frac{1}{8}\sqrt{67}$. In particular,

$$\sqrt{1 + \frac{3}{64}} = 1 + \frac{1}{2}\left(\frac{3}{64}\right) - \frac{1}{8}\left(\frac{3}{64}\right)^3 + R_3\left(\frac{3}{64}\right)$$

$$\frac{1}{8}\sqrt{67} = 1 + .0234375 - .0000128 + R_3\left(\frac{3}{64}\right)$$

Then $\sqrt{67} = 8.1876024 + 8\,R_3(\frac{3}{64})$, which indicates the remainder term is already very small, even for the value of $n = 3$. $\square$

We conclude this section with a comment to show that at least part of the conclusion of Taylor's Theorem is very intuitive. For if we suppose that a function f can be written in the form

$$f(x) = a_0 + a_1(x - a) + a_2(x - a)^2 + \cdots + a_n(x - a)^n + \cdots$$

then by replacing x by a, we obtain $f(a) = a_0$.

If we next differentiate f and set $x = a$, we obtain

$$f'(x) = a_1 + 2a_2(x - a) + 3a_3(x - a)^2 + \cdots$$

so that $f'(a) = a_1$.

By continuing this process, we find $f''(a) = 2a_2$, $f'''(a) = 6a_3$, and in general, $f^{(n)}(a) = n!\,a_n$. This value of

$$a_n = \frac{f^{(n)}(a)}{n!}$$

is the expression for the coefficients that occur in Taylor's Theorem. The part we cannot obtain from this approach is the remainder term $R_n(x)$, and this term is needed to show that a function f does have an expansion of the form $\sum_{k=0}^{\infty} a_k (x-a)^k$. More is said about this in Chapter 9.

EXERCISES 21

1. Use Taylor's Theorem with $n = 2$ to find upper and lower bounds for $\sqrt{67}$ Refer to Example 1.

2. Use Taylor's Theorem with $n = 4$ to find upper and lower bounds for $\ln .5$ Refer to Exercise 20.5c.

3. Use Taylor's Theorem with $n = 5$ and $a = 0$ to show that

$$\left| (\sin x) - \left(x - \frac{x^3}{6} + \frac{x^5}{120} \right) \right| < \frac{1}{720} \quad \text{for } |x| \leq 1$$

and then use Taylor's Theorem with $n = 6$ and $a = 0$ to show that

$$\left| (\sin x) - \left(x - \frac{x^3}{6} + \frac{x^5}{120} \right) \right| < \frac{1}{5040} \quad \text{for } |x| \leq 1$$

4. (a) Find the remainder term R_n for $f(x) = e^x$ with $a = 0$. Refer to Exercise 19.13c.

   (b) Show that $\lim_{n \to \infty} R_n = 0$ for all x.

5. (a) Find the remainder term R_n for $f(x) = \sqrt{x}$ with $a = 1$. Refer to Exercise 19.13d.

   (b) For which values of x will $\lim_{n \to \infty} R_n = 0$?

6. Go through the following steps to prove Taylor's Theorem for the case where $n = 2$. Assume that $f(t)$, $f'(t)$ and $f''(t)$ are all continuous on $[a, b]$, and $f'''(t)$ exists on (a, x).

   (a) Define $F(t) = f(x) - f(t) - (x-t)f'(t) - \frac{(x-t)^2}{2!} f''(t)$. Prove that $F'(t) = -\frac{(x-t)^2}{2!} f'''(t)$.

   (b) Define $G(t) = F(t) - \left(\frac{x-t}{x-a} \right)^3 F(a)$. Prove that $G(a) = 0$ and $G(x) = 0$.

   (c) By Rolle's Theorem, $G'(c) = 0$ for some $c \in (a, x)$. Substitute in the values for $F(a)$ and $F'(c)$ and show that this expression simplifies to

$$f(x) = f(a) + f'(a)(x-a) + \frac{f''(a)}{2!}(x-a)^2 + \frac{f'''(c)}{3!}(x-a)^3$$

7. Provide the missing details for the proof of Theorem 21.1.

SECTION 22   L'HOSPITAL'S RULE

Historical comment.   Section 18 contained comments about the role of Newton and Leibniz in the development of the derivative concept and the role of John and James Bernoulli, among others, in carrying out the details of this development. Newton's reluctance to publish and Leibniz's unclear writings made early progress difficult, but the excitement generated by the power of these new techniques was sufficient to overcome such obstacles. Leonhard Euler learned his calculus from John Bernoulli, and in 1748 published his two-volume work *Introductio in Analysin Infinitorum*, which forged the new branch of mathematics we call analysis. It is a very readable work (when translated from the Latin) and seems modern because Euler introduced a large portion of what is now current notation and terminology. Springer-Verlag published in 1988 an English translation of Book 1 of this work. Euler wrote more than 75 volumes of mathematics during his long lifetime, including the *Institutiones Calculi Differentialis* in 1755 and the three-volume *Institutiones Calculi Integralis* in 1768–1770.

The very first calculus text was written by another student of John Bernoulli, one who was far less gifted mathematically than Euler. The Marquis de L'Hospital paid Bernoulli a regular salary, and in return, Bernoulli agreed to provide instruction in the new calculus and also to communicate his new discoveries to the Marquis, who could do whatever he wished with them. L'Hospital chose to publish these findings under his own name in 1696 as the first differential calculus textbook. Of course, Bernoulli did not like this, but he had agreed to this arrangement. This text is best known for the result discovered by John Bernoulli, which is now known as L'Hospital's Rule, namely that if f and g are differentiable functions on $[a, b]$ with $f(a) = g(a) = 0$, then $\lim_{x \to a} \frac{f(x)}{g(x)} = \frac{f'(a)}{g'(a)}$. L'Hospital's proof was given verbally without functional notation, but the essence of his proof was that

$$\frac{f(a + dx)}{g(a + dx)} = \frac{f(a) + f'(a)dx}{g(a) + g'(a)dx} = \frac{f'(a)\,dx}{g'(a)\,dx} = \frac{f'(a)}{g'(a)}$$

We next demonstrate how to construct a modern proof using the limit concept. There are several cases to consider in establishing a complete proof. Since the details are often more tedious than instructive, we will state all the necessary results, but will prove only two cases and suggest

references where proofs for the others may be found. It may be helpful
to think of L'Hospital's Rule as providing answers to several of the cases
of the general limit theorem for functions as presented in Section 11. We
first need a result that is a generalization of the Mean Value Theorem, one
that involves two functions.

THEOREM 22.1 (Cauchy Mean Value Theorem)

Let f and g be continuous on $[a, b]$ and differentiable on (a, b), and assume
$g'(x) \neq 0$ for $a < x < b$. Then there is some $c \in (a, b)$ so that

$$\frac{f'(c)}{g'(c)} = \frac{f(b) - f(a)}{g(b) - g(a)}$$

Proof.   The proof follows immediately by applying Rolle's Theorem to
the function

$$h(x) = \frac{f(b) - f(a)}{g(b) - g(a)} \Big(g(x) - g(a)\Big) - \Big(f(x) - f(a)\Big)$$

The details are left as an exercise. It may be of interest to observe that
the Mean Value Theorem is obtained as a special case when $g(x) = x$.   ∎

We begin with a theorem whose proof closely resembles in its form, the
one presented in L'Hospital's book almost three centuries ago.

THEOREM 22.2   Let f and g be defined on the interval $[a, b)$ with
$f(a) = g(a) = 0$, and assume $g(x) \neq 0$ for $a < x < b$. If $f'(a)$ and $g'(a)$
exist and $g'(a) \neq 0$, then

$$\lim_{x \to a^+} \frac{f(x)}{g(x)} \quad \text{exists and equals} \quad \frac{f'(a)}{g'(a)}$$

Proof.   This is left as an exercise.

Comment.   Because many limits of the $\frac{0}{0}$ form do not meet the conditions of Theorem 22.2, we generalize the hypotheses in order to obtain a
theorem that is more widely applicable. Such a list of hypotheses usually
includes the following.

(1) f and g are continuous on the closed interval $[a, b]$.

(2) f and g are differentiable on the open interval (a, b).

(3) $\lim_{x \to a+} f(x) = 0$ and $\lim_{x \to a+} g(x) = 0$.

(4) $f(a) = g(a) = 0$.

(5) $g'(x) \neq 0$ for $x \in (a, b)$.

(6) $g(x) \neq 0$ for $x \in (a, b)$.

We need conditions (1) and (2) in order to apply the Cauchy Mean Value Theorem. We need conditions (5) and (6) to guarantee that the quotient functions $\frac{f(x)}{g(x)}$ and $\frac{f'(x)}{g'(x)}$ are well-defined. Condition (6) need not be stated separately since it is implied by the others. For if $g(x) = 0$ for some $x \in (a, b)$, since $g(a) = 0$, Rolle's Theorem would require that $g'(c) = 0$ for some $c \in (a, x)$, which is contrary to condition (5). Similarly, condition (4) is implied by conditions (1) and (3).

Condition (3) is needed to obtain the crucial relationship between

$$\lim_{x \to a+} \frac{f(x)}{g(x)} \quad \text{and} \quad \lim_{x \to a+} \frac{f'(x)}{g'(x)}$$

We are now ready to state and prove the main form of L'Hospital's Rule. We will prove statement (1) using a sequence argument, statement (2) by an epsilon argument, and leave statement (3) as an exercise. Results with very similar proofs are true for the cases of a left-hand limit where $x \to b^-$, and as x approaches $+\infty$ or $-\infty$.

THEOREM 22.3   Let f and g be continuous on $[a, b]$ and differentiable on (a, b). Assume also that $\lim_{x \to a+} f(x) = 0$, $\lim_{x \to a+} g(x) = 0$ and $g'(x) \neq 0$ for $a < x < b$.

(1) If $\lim\limits_{x \to a+} \dfrac{f'(x)}{g'(x)} = L \in \mathbf{R}$, then $\lim\limits_{x \to a+} \dfrac{f(x)}{g(x)} = L$

(2) If $\lim\limits_{x \to a+} \dfrac{f'(x)}{g'(x)} = +\infty$, then $\lim\limits_{x \to a+} \dfrac{f(x)}{g(x)} = +\infty$

(3) If $\lim\limits_{x \to a+} \dfrac{f'(x)}{g'(x)} = -\infty$, then $\lim\limits_{x \to a+} \dfrac{f(x)}{g(x)} = -\infty$

Proof of (1).   Let $\{x_n\}$ be any sequence so that $a < x_n < b$ and $\lim_{n \to \infty} x_n = a$. We must show that $\lim_{n \to \infty} \left(\frac{f}{g}\right)(x_n) = L$. For each n, by applying the Cauchy Mean Value Theorem to f and g on $[a, x_n]$, there is some $c_n \in (a, x_n)$ for which

$$\frac{f'(c_n)}{g'(c_n)} = \frac{f(x_n) - f(a)}{g(x_n) - g(a)} = \frac{f(x_n)}{g(x_n)}$$

Thus $\{c_n\}$ is a sequence so that $a < c_n < x_n < b$, $\lim_{n \to \infty} c_n = a$, and $\lim_{n \to \infty} \frac{f'(c_n)}{g'(c_n)} = L$. But then

$$\lim_{n \to \infty} \frac{f(x_n)}{g(x_n)} = \lim_{n \to \infty} \frac{f'(c_n)}{g'(c_n)} = L$$

and the proof is complete. ∎

Proof of (2).   Let $K > 0$ be given. We must find $\delta > 0$ so that if $a < x < a + \delta$, then $\frac{f(x)}{g(x)} > K$. Since $\lim_{x \to a+} \frac{f'(x)}{g'(x)} = +\infty$, there is $\delta > 0$ so that if $a < c < a + \delta$, then $\frac{f'(c)}{g'(c)} > K$. Choose x so that $x \in (a, a + \delta)$. By applying the Cauchy Mean Value Theorem to f and g on $[a, x]$, there is some $c \in (a, x)$ so that

$$\frac{f'(c)}{g'(c)} = \frac{f(x) - f(a)}{g(x) - g(a)} = \frac{f(x)}{g(x)}$$

and therefore $\frac{f(x)}{g(x)} > K$ as needed.   ∎

Proof of (3).   This is left as an exercise.

Example 1   Use L'Hospital's Rule to find $\lim_{x \to 0+} \left(\frac{1}{x} - \cot x \right)$.

Solution.   We first use algebra to show $\frac{1}{x} - \cot x = \frac{\sin x - x \cos x}{x \sin x}$.   Then by the use of Theorem 22.3,

$$\lim_{x \to 0+} \left(\frac{1}{x} - \cot x \right) = \lim_{x \to 0+} \frac{\sin x - x \cos x}{x \sin x}$$

$$= \lim_{x \to 0+} \frac{x \sin x}{x \cos x + \sin x}$$

$$= \lim_{x \to 0+} \frac{x \cos x + \sin x}{2 \cos x - x \sin x}$$

$$= 0 \quad \square$$

Example 2   Show that we cannot use L'Hospital's Rule to evaluate

$$\lim_{x \to 0+} \frac{x^2 \sin \frac{1}{x}}{\sin x}$$

Solution.   When we apply Theorem 22.3, we obtain

$$\lim_{x \to 0+} \frac{x^2 \sin \frac{1}{x}}{\sin x} = \lim_{x \to 0+} \frac{2x \sin \frac{1}{x} - \cos \frac{1}{x}}{\cos x}$$

However, the second limit does not exist because $\lim_{x \to 0+} \cos \frac{1}{x}$ does not exist, and so L'Hospital's Rule is inconclusive. We are able to evaluate this limit by use of the Limit Theorem because

$$\lim_{x \to 0+} \frac{x^2 \sin \frac{1}{x}}{\sin x} = \lim_{x \to 0+} \left(\frac{x}{\sin x} \right) \lim_{x \to 0+} \left(x \sin \frac{1}{x} \right) = 1 \cdot 0 = 0 \quad \square$$

A result similar to Theorem 22.3 can be stated and proved for left-handed limits as $x \to b^-$, and this is left as an exercise. We next consider the case where $x \to +\infty$ or $x \to -\infty$.

THEOREM 22.4   Let f and g be differentiable (and hence continuous) on the interval $[b, \infty)$. Assume that $\lim_{x \to \infty} f(x) = 0$, $\lim_{x \to \infty} g(x) = 0$ and $g'(x) \neq 0$ for all $x > b$ (we may also assume that $g(x) \neq 0$ for $x > b$ as well). If $\lim_{x \to \infty} \frac{f'(x)}{g'(x)} = \alpha$, (where α can be finite or infinite), then $\lim_{x \to \infty} \frac{f(x)}{g(x)} = \alpha$ also.

Proof.   In order to apply the results in Theorem 22.3, we use the change of variable $t = \frac{1}{x}$ to define the functions F and G on the interval $[0, \frac{1}{b}]$ by

$$F(t) = \begin{cases} f(\frac{1}{t}) & \text{if } 0 < t < \frac{1}{b} \\ 0 & \text{if } t = 0 \end{cases}$$

$$G(t) = \begin{cases} g(\frac{1}{t}) & \text{if } 0 < t < \frac{1}{b} \\ 0 & \text{if } t = 0 \end{cases}$$

Since by the Chain Rule, $F'(t) = f'(t)(-\frac{1}{t^2})$, we have

$$\lim_{t \to 0+} \frac{F'(t)}{G'(t)} = \lim_{t \to 0+} \frac{f'(t)\left(-\frac{1}{t^2}\right)}{g'(t)\left(-\frac{1}{t^2}\right)} = \lim_{x \to \infty} \frac{f'(x)}{g'(x)} = \alpha$$

Then by Theorem 22.3,

$$\lim_{t\to 0^+} \frac{F(t)}{G(t)} = \lim_{t\to 0^+} \frac{F'(t)}{G'(t)} = \alpha$$

and finally

$$\lim_{x\to\infty} \frac{f(x)}{g(x)} = \lim_{t\to 0^+} \frac{F(t)}{G(t)} = \alpha \quad \blacksquare$$

Example 3   Use Theorem 22.4 to evaluate $\lim_{x\to\infty}\left(1+\frac{1}{x}\right)^x$.

Solution.   We begin with the observation that

$$\left(1 + \frac{1}{x}\right)^x = e^{\ln\left(1+\frac{1}{x}\right)^x} = e^{x\,\ln\left(1+\frac{1}{x}\right)}$$

Then by using Theorem 22.4 and the continuity of e^x,

$$\lim_{x\to\infty} x\ln\left(1+\frac{1}{x}\right) = \lim_{x\to\infty} \frac{\ln\left(1+\frac{1}{x}\right)}{\frac{1}{x}} = \lim_{x\to\infty} \frac{1}{1+\frac{1}{x}} = 1$$

and therefore $\lim_{x\to\infty}\left(1+\frac{1}{x}\right)^x = e^1$. $\square$

The second indeterminate form is $\frac{\infty}{\infty}$, and while the same general result holds, namely $\lim \frac{f(x)}{g(x)} = \lim \frac{f'(x)}{g'(x)}$, the proof is more complicated.

THEOREM 22.5   Let f and g be differentiable on (a,b) and $g'(x) \neq 0$ for all $x \in (a,b)$. Assume also that $\lim_{x\to a^+} f(x) = \lim_{x\to a^+} g(x) = +\infty$. If $\lim_{x\to a^+} \frac{f'(x)}{g'(x)} = \alpha$ when α is finite, $+\infty$, or $-\infty$, then $\lim_{x\to a^+} \frac{f(x)}{g(x)} = \alpha$ also.

Comment.   After seeing the method used in the proof of Theorem 22.4 in order to make use of Theorem 22.3, you might be tempted to try a similar substitution here. If we define $F(x) = \frac{1}{f(x)}$ and $G(x) = \frac{1}{g(x)}$, then

$$\lim_{x\to a^+} \frac{f(x)}{g(x)} = \lim_{x\to a^+} \frac{G(x)}{F(x)}$$

and since $\lim_{x \to a+} \frac{G(x)}{F(x)}$ is of the $\frac{0}{0}$ form, we could try to apply Theorem 22.3 to the functions F and G. Can you determine where this approach will run into trouble?

A correct proof for the $\frac{\infty}{\infty}$ form of L'Hospital's Rule is considerably more involved than the one for the $\frac{0}{0}$ form, and we choose to omit the details here. The interested reader can find a proof using an epsilon argument[1] or a proof using a geometric approach.[2] Complete proofs must not only treat the separate cases when α is finite, $\alpha = +\infty$, or $\alpha = -\infty$, but must also consider the possibilities of a left-hand limit $(x \to b^-)$ or an infinite limit $(x \to +\infty$ or $x \to -\infty)$. The theorem needs to be shown valid when $\lim |f(x)| = +\infty$ and $\lim |g(x)| = +\infty$ (one case is considered as an exercise). So while L'Hospital's Rule is undeniably a useful, perhaps indispensable, tool in analysis, a complete proof of all the possible cases requires a great deal of rather tedious effort. That is why we chose to only consider a few of the most representative cases.

It is possible to apply these theorems to the remaining five indeterminate forms $(0 \cdot \infty, \ \infty - \infty, \ 1^\infty, \ 0^0, \ \infty^0)$ by the use of appropriate algebraic methods, as was illustrated in Examples 1 and 3.

Example 4   Apply Theorem 22.5 to evaluate $\lim_{x \to 0+} \frac{-\ln x}{\csc x}$.

Solution.   We should first verify that all the hypotheses of Theorem 22.5 are satisfied for $f(x) = -\ln x$ and $g(x) = \csc x$ on the interval $(0, 1)$. Then

$$\lim_{x \to 0+} \frac{-\ln x}{\csc x} = \lim_{x \to 0+} \frac{1}{x \csc x \cot x} = \lim_{x \to 0+} \frac{\sin^2 x}{x \cos x}$$

$$= \lim_{x \to 0+} \frac{\sin x}{x} \lim_{x \to 0+} \tan x = 0 \quad \square$$

---

[1] *Introduction to Real Analysis* by Robert Bartle and Donald Sherbert, New York: John Wiley and Sons, pp. 217–218.
[2] *Mathematical Analysis, Second Edition* by Thomas Apostle, Reading, MA: Addison-Wesley, pp. 406–408.

EXERCISES 22

1. Use L'Hospital's Rule to evaluate the following limits. Identify which theorem you are using.

   (a) $\lim_{x \to 0+} \frac{\tan x - x}{x^3}$

   (b) $\lim_{x \to 0+} \frac{10^{\sqrt{x}} - 1}{2^{\sqrt{x}} - 1}$

   (c) $\lim_{x \to \frac{\pi}{2}-} (\sec x - \tan x)$

   (d) $\lim_{x \to \infty} \frac{x}{3 \ln x}$

   (e) $\lim_{x \to \infty} x \ln \left(1 + \frac{3}{x}\right)$

   (f) $\lim_{x \to 0+} \frac{1}{x (\ln x)^2}$

2. Explain why L'Hospital's Rule does not apply to each of the following limits.

   (a) $\lim_{x \to 0} \frac{\cos x}{x^2}$

   (b) $\lim_{x \to 0} (\sec 2x)^{x^2}$

   (c) $\lim_{x \to \infty} \frac{x - \sin x}{x}$

3. Use an appropriate algebraic identity so that each of the following limits can be evaluated by one of the forms of L'Hospital's Rule in this section.

   (a) $\lim_{x \to \infty} \left(1 + \frac{a}{x}\right)^{bx}$

   (b) $\lim_{x \to 0} (\cos x)^{\frac{1}{x^2}}$

   (c) $\lim_{x \to 0+} \left(\frac{1}{x} - \frac{1}{\arctan x}\right)$

4. Show that $\lim_{t \to \infty} \frac{t^x}{e^t} = 0$ for all values of x. For which values of x do you need to use L'Hospital's Rule?

5. Sketch a graph of the function $f(x) = x^{\frac{1}{x}}$, indicating asymptotes and relative maxima or minima. Use L'Hospital's Rule as needed.

6. Prove Theorem 22.2.

7. Prove statement (3) of Theorem 22.3.

8. Prove statement (2) of Theorem 22.3 for the case where the limit is taken as $x \to b^-$.

9. Why is Example 2 not a contradiction of Theorem 22.3?

6

The Riemann Integral

SECTION 23 A HISTORICAL INTRODUCTION
TO THE INTEGRAL

There have been three clearly defined periods in the development of the integral concept, coinciding with the three periods in the historical development of calculus itself, as described in Section 3. The first period consists of the discovery of methods, often very ingenious ones, for finding areas and volumes of specific regions and solids. Archimedes in the Greek period obtained many familiar formulas, such as the ones for the area of a circle and ellipse, and the volume and surface area for a sphere. After a gap of almost two thousand years, several European mathematicians in the seventeenth century began to again consider this problem of areas and volumes, and were able to expand on the results and methods of Archimedes. Some of their accomplishments will be discussed in this section.

The second period was ushered in when Newton and Leibniz discovered that the lengthy summing process used in the first period could be replaced by the simpler technique of finding an antiderivative for a specified function. Although the definite integral is introduced in a calculus course by using sums of rectangular areas to approximate the desired area, as soon as the above result known as the Fundamental Theorem of Calculus is presented, integration and antidifferentiation are forever identified in the student's mind. There is really no shame in this, since this was exactly how all mathematicians viewed integration during the second period, from the time of the discovery of the calculus around 1670 until the time of Cauchy around 1820. This way of thinking was sufficient for the discovery of a remarkable range of applications, including motion of particles (especially the planets), the mechanics of rods and fibers, the shape of hulls and sails for ships, and results in optics and heat.

The third period, characterized by a more theoretical approach, became necessary when the simpler concept of antidifferentiation was unable to handle the new kinds of discontinuous functions, which came in through the work of Joseph Fourier on the theory of heat. Cauchy applied his new definition of limit to define the definite integral of the function f defined on $[a, b]$ as the limit as n gets very large (i.e. approaches infinity) of sums of the form $\sum_{i=1}^{n} f(x_i)(x_i - x_{i-1})$, where the collection of points $\{x_i\}_{i=0}^{n}$ divides the interval $[a, b]$ into n parts. He then used this definition to derive many basic properties of the definite integral and to provide the first rigorous proof for the Fundamental Theorem. Other mathematicians, such as Darboux, Riemann, Stieltjes, and Lebesgue offered refinements and extensions to this definition.

We now provide a sample of the summation of area problems that were typical of the first period. It is not only very interesting material, but it introduces you to many of the formulas and techniques that are used in a modern treatment.

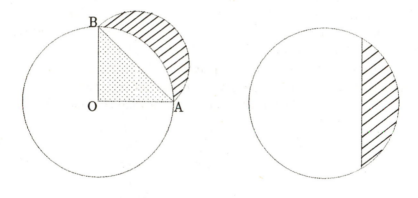

Figure 23.1   Figure 23.2

We begin with a circle whose radius is the length of the line segment OA, and then construct a semicircle with line segment AB as a diameter (see Figure 23.1). The portion of the semicircle that is outside the first circle was called a "lunar crescent" by the Greeks. Hippocrates, who was a contemporary of Zeno in the fifth century B.C., proved that the area of this lunar crescent is equal to the area of the right triangle AOB.[1] This was the first successful attempt to show that a curvilinear region

---

[1] *Mathematical Thought From Ancient To Modern Times* by Morris Kline, New York: Oxford University Press, 1972, p. 41.

can be "commensurable" with a straight-sided figure, which means that a region bounded by curved lines has the same area as a region bounded by straight lines, such as a triangle, square, or polygon. It was thus a part of the "squaring the circle" problem, one of the three famous construction problems proposed by the Greeks, but not solved until the nineteenth century. The Greeks also worked unsuccessfully on the problem to find the area of a part of a circle cut off by a straight line segment (see Figure 23.2).

The method of Hippocrates could not be extended to find either the area of the entire circle or a part cut off by a straight line segment. Archimedes tackled these problems two centuries later and also could not solve either one. He used the technique of inscribed and circumscribed polygons to approximate the area of the circle, but although he used up to 96-sided polygons, he was only able to obtain two-decimal-place accuracy. No matter how many more sides he might have used however, he would never get an exact answer. Neither Archimedes nor others could solve the quadrature of the ellipse or hyperbola, but Archimedes was able to solve the quadrature of the parabola by showing that the area of a parabolic segment is equal to a rational multiple (namely $\frac{4}{3}$) of the area of an associated triangle[1] (see Figure 23.3). Archimedes actually provided three proofs for this result, which are of interest for their diversity and creativity.

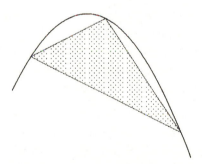

Figure 23.3

The first method of proof uses polygons inscribed within the parabolic segment, showing at each stage that the new area added is one-quarter that of the previous area. A geometric series is obtained, whose value

---

[1] *The Historical Development of the Calculus* by C. H. Edwards, Jr, Springer-Verlag, 1979, pp. 35-39.

Archimedes was able to calculate. Instead of the limit procedure that we use today, Archimedes used the method of exhaustion. This technique resembles mathematical induction, in that you need to know the answer before you can apply the method to verify the result. Therefore it is a method of proof but not a means of discovery. The method of exhaustion is a proof by contradiction in that one proves $A = B$ by showing that $A < B$ or $A > B$ cannot hold.

The second method Archimedes used to solve the quadrature of the parabola problem is most like the modern approach of approximating sums. The use of rectangles distinguishes the second method from the polygon approach. Instead of a geometric series that can be easily summed, the sum $\sum_1^n i^2$ is obtained instead. Fortunately, Archimedes knew this sum equalled $\frac{1}{6}n(n+1)(2n+1)$, so he once again was able to find the answer.

The third approach was actually called *The Method* after a book written by Archimedes with this title. A copy of this book was accidently found in 1906, having been erased and written over during the medieval period. Archimedes said that this physical approach is the method he used to obtain many of his results, and he describes his thought processes when attacking a problem. He did not accept this method as providing a proof, but once he knew the answer, he could use more rigorous means to establish the result. Using the idea of weights and levers, and by considering the area of the parabolic segment as consisting of its straight line cross-sections, Archimedes located the fulcrum so that the area of the parabolic segment concentrated at its center of gravity would balance the area of an associated triangle concentrated at another point. The technique uses proportions and parabolic identities. Cavalieri would later also think of area as made up of straight line cross-sections.

After Archimedes, no one made further progress with this question of area and volume until the early seventeenth century when several individuals before Newton and Leibniz achieved success for a number of particular cases. We consider briefly the contributions of Kepler, Cavalieri, Fermat, and Gregory.

Johann Kepler is best known for his three laws of planetary motion, the second of which deals with areas, namely that the radius vector from the sun to a planet sweeps out area at a constant rate. We should not conclude from this that Kepler used integration techniques to establish this result. He discovered the result empirically from the wealth of observational data that he had available, and his proof worked out only because the errors involved in his work compensated one another in a fortuitous way. However, Kepler was familiar with the idea of dissection into an "infinite number of infinitesmal pieces," and he used this idea in a very practical way to find the volumes of more than 90 solids of revolution.

Cavalieri, who was an Italian disciple of Galileo, published a book in 1635 that presented a systematic approach to the use of infinitesimals. His writings encouraged others to study this subject. He considered areas as composed of lines, and volumes as composed of areas, as Archimedes did before him. Cavalieri's Theorem states that if two solids have equal altitudes and if sections made by planes parallel to the bases and at equal distances from them are always in a given ratio, then the volumes of the solids are also in this ratio. He used this idea to correctly find the area of an ellipse, given the area of a circle. This method can lead to errors and sometimes gives incorrect formulas.

Cavalieri was more successful when he generalized Archimedes' solution for the quadrature of the parabola by his second method. In particular, Cavalieri showed that the area under the curve $y = x^3$ between 0 and 1 could be expressed by the use of n approximating rectangles with bases of equal length in terms of the sum $\sum_1^n i^3$. But this sum could be written as $\frac{1}{4}n^2(n + 1)^2$ according to a result known by the Arabs, who used a geometric construction approach. Encouraged by this success, Cavalieri found that the area under $y = x^k$ between $x = 0$ and $x = 1$ could be expressed in terms of $\sum_1^n i^k$ for $k = 4, 5, 6, 7, 8, 9$ and that the area was $\frac{1}{k+1}$ in each case. The necessary calculations became increasingly long, and Cavalieri gave up on the problem as unsolvable for general values of k.

Knowing that a geometric series was much easier to sum than was $\sum_1^n i^k$, Fermat had the insight around 1650 to partition the interval $[0, 1]$ into n unequal parts, using the points $0 < p^n < \cdots < p^2 < p < 1$, and then to sum the areas of the n rectangles erected under the curve using these bases. This sum leads to a geometric series that quickly yields an area of $\frac{1}{k+1}$ for the region under the curve $y = x^k$ and above the interval $[0, 1]$ on the x-axis. In Fermat's method, k could be any positive or negative rational exponent, with the one exception of $k = -1$. This blending of the first two methods of Archimedes provided a basis for further advances in the problem of area calculation.

James Gregory of Scotland took up the remaining question of the area under the hyperbola $y = \frac{1}{x}$ and was able to use yet another means for partitioning the base interval in order to show that this area function has the properties of a logarithm. Gregory's writings were based on geometry and were often difficult to follow. On the other hand, John Wallis of England published *Arithmetica Infinitorum* in 1655, which used the new algebraic techniques instead of the older and more cumbersome geometric techniques. This gave calculus results in the best form possible before the contributions of Newton and Leibniz.

Whereas many individuals before Newton and Leibniz established important area and volume results using a variety of methods, none of them apparently saw how to develop a general method to solve all these related problems. This is what Newton and Leibniz achieved, and this is why their contributions were so much more valuable, essentially rendering everyone else's work obsolete. The notation that they developed (especially the differentials of Leibniz), the development of algorithmic techniques for calculating derivatives to find slopes of tangent lines, and the use of anti-derivatives to find areas and volumes, constitute what we often refer to as the discovery of calculus. These discoveries provided a basis for their successors in the next century to build upon, and allowed them to achieve remarkable success in answering questions about the physical world.

SECTION 24   DEFINITION OF THE INTEGRAL

In Chapter 5, we were able to give the definition of the derivative in just a single phrase—f is differentiable at c if $\lim_{x \to c} \frac{f(x)-f(c)}{x-c}$ exists. The situation is quite different for the definition of the definite integral. It requires a fairly long section to present all the notation and concepts that are needed. In spite of this complexity, you should remember that area under a curve remains the motivating idea underlying all the symbolism.

Definition 1   A *partition P* of the closed interval $[a, b]$ is a finite collection of points $\{x_i\}_{i=0}^n$ so that $a = x_0 < x_1 < x_2 < \cdots < x_n = b$. The *norm of the partition P* is defined by $|P| = \max\{\Delta x_i : i = 1, \ldots, n\}$, where $\Delta x_i = x_i - x_{i-1}$. The collection of points $\{c_i\}_{i=1}^n$ will be said to *belong to P* if $x_{i-1} \le c_i \le x_i$ for $i = 1, \ldots, n$.

The collection of right endpoints $\{x_i\}_{i=1}^n$ is a common choice for the c_i values; the collection of midpoints $c_i = \frac{x_i + x_{i-1}}{2}$ is another possibility.

Definition 2   If $P = \{x_i\}_{i=0}^n$ is a partition of $[a, b]$, then the partition P^* will be called a *refinement of P* if P^* contains all the points of P along with a finite number of additional points from the interval $[a, b]$. We could also express this by saying that P is a subset of P^*.

Historical Comment.   The third period in the development of the integral concept was introduced through the publication in 1823 of the lectures of Cauchy, which were given at the Ecole Polytechnique in Paris. Though his notation is somewhat different than what we are used to, it is surprisingly modern. We amend his notation somewhat to make it easier to follow, but try to preserve his basic ideas and approach. For each partition $P = \{x_i\}_{i=0}^n$ of $[a, b]$, he considered the approximating sum $S = \sum_{i=1}^n f(x_{i-1})(x_i - x_{i-1})$, and defined the integral $\int_a^b f(x)\, dx$ as the limit of these sums as the length of the subintervals approaches zero. In Cauchy's words,

> when the elements of the partition of $[a, b]$ become infinitely small, the mode of division has no more than an imperceptible influence on the value of S; and, if one makes the numerical values of these elements decrease indefinitely by increasing their

number, the value of S will end by being perceptibly constant, or in other words, it will end by attaining a certain limit which will depend solely on the form of the function f and on the extreme values a and b attributed to the variable x. This limit is that which one calls a definite integral.[1]

The form in which this definition is expressed is much different from what we see in most calculus texts, in that it relies more on words than symbols, but Cauchy made a great stride in separating the definition of the integral from the antiderivative concept, and in expressing it as a limit of an approximating sum. In this form, it was ready to be refined by the successors of Cauchy into a modern tool of great generality and power.

It was George Bernard Riemann who adjusted Cauchy's integral definition to its modern form by permitting the approximating sums to be determined not by just the left endpoint of each subinterval of a partition, but by <u>any</u> point $c_i \in [x_{i-1}, x_i]$.

Definition 3   For a function f bounded on $[a, b]$, a partition $P = \{x_i\}_{i=0}^{n}$ of $[a, b]$, and a collection of points $\{c_i\}_{i=1}^{n}$ belonging to P, the finite sum $\sum_{i=1}^{n} f(c_i)\,\Delta x_i$ is called the *Riemann sum* determined by $\left(f, P, \{c_i\}\right)$.

It is traditional in calculus to define the definite integral $\int_a^b f(x)\,dx$ in terms of the limit expressed as $\lim_{|P| \to 0} \sum_{i=1}^{n} f(c_i)\,\Delta x_i$ (see Definition 4 below).

Definition 4   Let f be defined and bounded on $[a, b]$. Then f is *Darboux integrable* on $[a, b]$ if there is a real number L so that for every $\epsilon > 0$, there is a $\delta_\epsilon > 0$, so that if $P = \{x_i\}_{i=0}^{n}$ is any partition of $[a, b]$ for which $|P| < \delta_\epsilon$ and $\{c_i\}_{i=1}^{n}$ is any collection of points belonging to P, then $\left| \left(\sum_{i=1}^{n} f(c_i)\Delta x_i \right) - L \right| < \epsilon$. In this case, L is called the value of the integral of f on $[a, b]$, and we write $L = \int_a^b f(x)\,dx$.

This definition has some weaknesses. It is too general to apply to a function that is not essentially constant, and it lacks a simple necessary and sufficient condition for integrability, which is essential for proving integral

---

[1] *The Historical Development of the Calculus* by C.H. Edwards, Jr.   New York: Springer-Verlag, 1979, p. 320.

properties and theorems. Riemann suggested how this might be remedied, and others carried out his ideas, which are expressed below in the integral definition that we use in our development of the integral.

Definition 5   Let f be defined and bounded on $[a, b]$ and let $P = \{x_i\}_{i=0}^n$ be a partition of $[a, b]$. Then define $m_i(f)$ and $M_i(f)$ by the formulas

$$m_i(f) \;=\; \inf\{f(x) : x \in [x_{i-1}, x_i]\}$$
$$M_i(f) \;=\; \sup\{f(x) : x \in [x_{i-1}, x_i]\}$$

When only one function is involved, we can shorten this notation to m_i and M_i. Then $\sum_{i=1}^n m_i \, \Delta x_i$ and $\sum_{i=1}^n M_i \, \Delta x_i$ are called the *lower sum* and *upper sum* respectively that are determined by the function f and the partition P. While a variety of notations are possible, we will use $L(f, P)$ to denote the lower sum and $U(f, P)$ to denote the upper sum.

It should be clear that if $\sum_{i=1}^n f(c_i) \, \Delta x_i$ is any Riemann sum determined by f and P, then

$$L(f, P) \;\leq\; \sum_{i=1}^n f(c_i) \, \Delta x_i \;\leq\; U(f, P)$$

For a continuous function f, $L(f, P)$ is the smallest Riemann sum, while $U(f, P)$ is the largest Riemann sum for a given f and P. Which theorem in Chapter 4 implies this result? It is left as an exercise to show that for an arbitrary function, $L(f, P)$ is the infimum of the set of all possible Riemann sums determined according to Definition 3, while $U(f, P)$ is the supremum of this set.

It is next necessary to establish the following three lemmas in order to express the definition of the Riemann integral.

LEMMA 24.1   If P^* is a refinement of the partition P of $[a, b]$, then for any function f defined on $[a, b]$,

$$L(f, P) \;\leq\; L(f, P^*) \;\leq\; U(f, P^*) \;\leq\; U(f, P)$$

Proof.   Assume first that $P = \{x_i\}_{i=0}^n$ is a partition of $[a, b]$ and $x^* \in P^*$ so that $x_{k-1} < x^* < x_k$. Then by Lemma 5.4,

$$\inf\{f(x) : x \in [x_{k-1}, x^*]\} \;\geq\; m_k$$

$$\inf\{f(x) : x \in [x^*, x_k]\} \;\geq\; m_k$$

It follows that

$$m_k \, \Delta x_k \;=\; m_k \, (x_k - x^* + x^* - x_{k-1})$$

$$=\; m_k \, (x_k - x^*) \;+\; m_k \, (x^* - x_{k-1})$$

$$\leq\; \inf \{ f(x) : [x^*, x_k] \}(x_k - x^*)$$

$$+\; \inf \{ f(x) : [x_{k-1}, x^*] \}(x^* - x_{k-1})$$

Since this inequality holds on any subinterval of the partition P to which one or more subdivision points are added in order to obtain the refinement P^*, we have

$$L(f, P) \;=\; \sum_{i=1}^{n} m_i \Delta x_i \;\leq\; L(f, P^*)$$

The shaded region in Figure 24.1 shows the value for the lower sum of f on the interval $[x_{k-1}, x_k]$ for the partition P, while the larger shaded region in Figure 24.2 shows the value for the lower sum of f on the interval $[x_{k-1}, x_k]$ for the partition P^*.

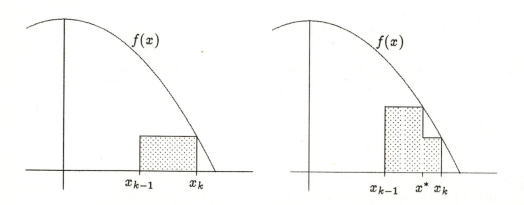

Figure 24.1   Figure 24.2

It is left as an exercise to show similarly that $U(f, P^*) \leq U(f, P)$. ∎

This lemma suggests that as more subdivisions are made, the lower sums increase and the upper sums decrease. Borrowing an idea from monotone sequences in Chapter 2 where an increasing sequence converges to $\sup\{x_n\}$ and a decreasing sequence converges to $\inf\{x_n\}$, we next define upper and lower integrals in terms of these upper and lower sums.

Definition 6 For a function f that is bounded on $[a, b]$, we define the *lower integral of f on $[a, b]$* as

$$\underline{\int_a^b} f = \sup\{L(f, P) : P \text{ is a partition of } [a, b]\}$$

and the *upper integral of f on $[a, b]$* as

$$\overline{\int_a^b} f = \inf\{U(f, P) : P \text{ is a partition of } [a, b]\}$$

These can be expressed more concisely by the symbols

$$\underline{\int_a^b} f = \sup_P L(f, P) \quad \text{and} \quad \overline{\int_a^b} f = \inf_P U(f, P)$$

LEMMA 24.2 If P_1 and P_2 are any two partitions of $[a, b]$ and f is any function defined and bounded on $[a, b]$, then $L(f, P_1) \leq U(f, P_2)$.

Proof. This result follows immediately from Lemma 24.1, when we observe that $P^* = P_1 \cup P_2$ is a refinement of both P_1 and P_2. Then

$$L(f, P_1) \leq L(f, P^*) \leq U(f, P^*) \leq U(f, P_2) \quad \blacksquare$$

LEMMA 24.3 For any function f defined and bounded on $[a, b]$, the upper and lower integrals exist, and $\underline{\int_a^b} f \leq \overline{\int_a^b} f$.

Proof. The desired inequality follows from some simple results listed in Section 5. We begin by choosing an arbitrary partition P_o of $[a, b]$. For any partition P of $[a, b]$, we have $L(f, P) \leq U(f, P_o)$ by Lemma 24.2. Then by Lemma 5.3,

$$\underline{\int_a^b} f = \sup_P L(f, P) \leq U(f, P_o)$$

By letting P_o next vary over all partitions of $[a, b]$, we have

$$\underline{\int_a^b} f \leq \inf_{P_o} U(f, P_o) = \overline{\int_a^b} f$$

once again by Lemma 5.3, and thus the proof is complete. ■

We come finally to the definition of the Riemann integral that will be used for the remainder of our work. See Exercise 14 for the reason why we restrict our consideration to bounded functions.

Definition 7   The bounded function f is called *Riemann integrable on* $[a, b]$, if the lower integral $\underline{\int_a^b} f$ and the upper integral $\overline{\int_a^b} f$ are finite and equal. In this case, $\int_a^b f$ is defined to be this common value.

Example 1   Apply Definition 7 to $f(x) = k$ on $[a, b]$.

Solution.   For any partition $P = \{x_i\}_{i=0}^n$, $m_i = M_i = k$ and so

$$L(f, P) = U(f, P) = \sum_{i=1}^n k \, \Delta x_i = k (b - a)$$

Also, $\underline{\int_a^b} f = \overline{\int_a^b} f = k(b-a)$, and so $f(x) = k$ is integrable on $[a, b]$, and has the value of $k(b-a)$.   □

Example 2   Apply Definition 7 to $f(x) = x$ on $[1, 4]$.

Solution.   For this increasing function, m_i occurs at the left endpoint, and M_i occurs at the right endpoint. Then $L(f, P) = \sum_{i=1}^n x_{i-1} \Delta x_i$ and $U(f, P) = \sum_{i=1}^n x_i \, \Delta x_i$, but this is as far as we can go with our work using Definition 7, even for such a simple function as $f(x) = x$. We will return to this problem in Example 5 after we obtain a few more results. □

Example 3   Apply Definition 7 to $f(x) = [x]$ on the interval $[1, 2]$.

Solution.   For any partition P of $[1, 2]$, $m_i = 1$ for all i, so $L(f, P) = \sum_{i=1}^n 1 \, \Delta x_i = 1 \cdot 1 = 1$, and therefore $\underline{\int_1^2} [x] \, dx = 1$.

Since $M_i = 1$, for $i < n$, and $M_n = 2$,

$$U(f, P) = \sum_{i=1}^{n-1} 1 \, \Delta x_i + 2 \, \Delta x_n = \sum_{i=1}^{n} 1 \, \Delta x_i + \Delta x_n = 1 + \Delta x_n$$

Since Δx_n can be made arbitrarily small, $\overline{\int_1^2} [x] \, dx = \inf (1 + \Delta x_n) = 1$, and so this function is Riemann integrable with value 1, even though it has a jump discontinuity at $x = 2$. □

Example 4   Apply Definition 7 to the function defined on the interval $[1, 3]$, by setting $f(x) = 1$ if x is rational, and $f(x) = 2$ if x is irrational.

Solution.   This example was first given by Dirichlet to show that a function could be defined with more than a finite number of discontinuities. It also provides the standard example of a bounded function that is not Riemann integrable, but which is Lebesgue integrable. The Lebesgue integral is a generalization of the Riemann integral, but it is not discussed in this book. For any partition P of $[1, 3]$, $m_i = 1$ and $M_i = 2$ for all i because there are rational and irrational numbers in any subinterval $[x_{i-1}, x_i]$, by the denseness property of these numbers. Then

$$L(f, P) = \sum_{i=1}^{n} 1 \, \Delta x_i = 1 \cdot 2 = 2$$

$$U(f, P) = \sum_{i=1}^{n} 2 \, \Delta x_i = 2 \cdot 2 = 4$$

which leads to $\underline{\int_1^3} f = 2$ and $\overline{\int_1^3} f = 4$. This function is not Riemann integrable because the upper and lower integral values are not equal. □

We next consider a special case of a Riemann sum, which is the one that is usually applied by students in calculus. It is also the form that was used by Archimedes and Cavalieri.

Special Case.   We use the symbol P_n to denote the partition of $[a, b]$ into n equal parts. In this case, all the Δx_i are equal to $\Delta x = \frac{b-a}{n}$ and also $x_i = a + i \Delta x$. Since the norm $|P_n| = \Delta x = \frac{b-a}{n}$, the limit as $|P_n| \to 0$ is equivalent to the limit as $n \to \infty$. We can now use the sequence concept to obtain another definition for $\int_a^b f(x) \, dx$.

Definition 8   For every natural number n, if we choose the partition P_n and the c_i values at the right endpoint x_i, then $S_n = \sum_{i=1}^{n} f(x_i)\Delta x_i$ will be one of the Riemann sum values. We say that f is *sequence integrable on* $[a, b]$ if the sequence $\{S_n\}$ converges, and define the value of the integral as $\int_a^b f(x)\,dx = \lim_{n\to\infty} S_n$ in this case.

This definition gives the same values as does Definition 7 for most of the familiar functions of calculus. See Example 4 for a counterexample. Most of the examples used in a calculus course to illustrate the integral definition use this form of the sequence integral.

The above examples should help you to better understand the nature of the integral definition. We are now ready to apply these ideas to obtain the integral theorems. The first theorem will be extremely useful because it provides a necessary and sufficient condition for a function to be Riemann integrable. We will use this result to prove most of the theorems in this chapter. It enables us to use the special partition P_n in our work instead of a general partition P .

THEOREM 24.4 (The Riemann Integrability Criterion)

The function f is Riemann integrable on $[a, b]$ if and only if for every $\epsilon > 0$, there is a partition P_ϵ of $[a, b]$, so that $U(f, P_\epsilon) - L(f, P_\epsilon) < \epsilon$.

Proof of $\Leftarrow$:   This result follows from the inequality that

$$L(f, P) \;\leq\; \underline{\int_a^b} f \;\leq\; \overline{\int_a^b} f \;\leq\; U(f, P)$$

for any partition P of $[a, b]$. Let $\epsilon > 0$ be given. Since there is a partition P_ϵ of $[a, b]$ for which $U(f, P_\epsilon) - L(f, P_\epsilon) < \epsilon$, it is also true that

$$0 \;\leq\; \overline{\int_a^b} f - \underline{\int_a^b} f \;<\; \epsilon$$

Since ϵ is arbitrary, $\underline{\int_a^b} f = \overline{\int_a^b} f$ and f is integrable on $[a, b]$.   ∎

Proof of $\Rightarrow$:   This result follows from the definition of the upper and lower integral in terms of an infimum and a supremum respectively.

Since $\int_a^b f = \inf_P U(f, P)$, there must be a partition P_1 of $[a, b]$, so that

$$\overline{\int_a^b} f < U(f, P_1) < \overline{\int_a^b} f + \frac{\epsilon}{2}$$

by Lemma 5.2. Similarly, since $\int_a^b f = \sup L(f, P)$, there must be a partition P_2 of $[a, b]$ so that

$$\underline{\int_a^b} f - \frac{\epsilon}{2} < L(f, P_2) < \underline{\int_a^b} f$$

Since $\underline{\int_a^b} f = \overline{\int_a^b} f = \int_a^b f$ and since $P_\epsilon = P_1 \cup P_2$ is a refinement both of P_1 and P_2, we have by Lemma 24.1 that

$$U(f, P_\epsilon) - L(f, P_\epsilon) < U(f, P_1) - L(f, P_2)$$

$$< \overline{\int_a^b} f + \frac{\epsilon}{2} - \underline{\int_a^b} f + \frac{\epsilon}{2} = \epsilon \quad \blacksquare$$

Example 5 Use Theorem 24.4 to show that $f(x) = x$ is integrable on $[1, 4]$. See Example 2.

Solution. Choose $n \in \mathbf{N}$ so that $\frac{9}{n} < \epsilon$, and let P_n be the partition of $[1, 4]$ into n equal parts. Observe that $\Delta x = \frac{3}{n}$ and $x_i = 1 + \frac{3i}{n}$, so that $M_i = f(x_i) = f(1 + \frac{3i}{n}) = 1 + \frac{3i}{n}$ and $m_i = f(x_{i-1}) = 1 + \frac{3(i-1)}{n}$. Then

$$U(f, P_n) - L(f, P_n) = \sum_{i=1}^{n} (M_i - m_i) \Delta x$$

$$= \sum_{i=1}^{n} \frac{3}{n} \cdot \frac{3}{n}$$

$$= \frac{9}{n^2} \sum_{i=1}^{n} 1 = \frac{9}{n} < \epsilon \quad \square$$

See Exercise 15 for a method to find the value of the integral $\int_1^4 x \, dx$.

EXERCISES 24

1. Apply the Riemann integral definition (Definition 7) to $f(x) = [x + 2]$ on the interval $[1, 2]$.

2. If $f(x) = x^2$ on $[2, 7]$, find $U(f, P_n)$.

3. (a) If $f(x) = x^2$ on $[0, b]$, find $U(f, P_n)$ and $L(f, P_n)$.

    (b) Show that f is integrable on $[0, b]$ by the use of Definition 7.

    (c) Show that f is integrable on $[0, b]$ by the use of Theorem 24.4.

4. If $f(x) = \sqrt{x}$ on $[0, 4]$, express $U(f, P_n)$ in simplest possible form.

5. Verify the statement after Definition 5 that $L(f, P)$ is the infimum of the set of all possible Riemann sums determined according to Definition 3, while $U(f, P)$ is the supremum of this set.

6. Let f be defined as in Example 4.

    (a) Find $\underline{\int_{\pi}^{\pi+2}} f$ and $\overline{\int_{\pi}^{\pi+2}} f$

    (b) Use Definition 8 to show that f is sequence integrable on $[\pi, \pi + 2]$. Compare the value of this integral with the one obtained in Example 4 for f on $[1, 3]$. What do you observe from this as a weakness of the sequence integral?

7. (a) Prove that $\sum_{i=1}^{n} i^p - (i - 1)^p = n^p$ for any positive integer p by the use of the telescoping sum method.

    (b) Show that $i^p - (i - 1)^p = \sum_{k=1}^{p}(-1)^{k+1}\binom{p}{k} i^{p-k}$ by use of the binomial expansion method.

8. Using the results in Exercise 7, we have the general formula that for all $p \in \mathbb{N}$,

$$\sum_{i=1}^{n} \sum_{k=1}^{p} (-1)^{k+1} \binom{p}{k} i^{p-k} = n^p$$

    (a) Write out this formula for the case where $p = 2$, and use this formula to show that $\sum_{i=1}^{n} i = \frac{n(n+1)}{2}$.

    (b) Write out this formula for the case where $p = 3$, and use this formula to show that $\sum_{i=1}^{n} i^2 = \frac{n(n+1)(2n+1)}{6}$.

    (c) Write out this formula for the case where $p = 4$, and use this formula to show that $\sum_{i=1}^{n} i^3 = \frac{n^2(n+1)^2}{4}$.

    (d) Look for the pattern in the above three cases and make a guess for the value of $\sum_{i=1}^{n} i^4$. Was your guess correct?

9. Let $f(x) = \frac{1}{x}$ be defined on $[1, 2]$.

(b) Show that $U(f, P_n) - L(f, P_n) = \frac{1}{2n}$ and that f is therefore integrable on the interval $[1, 2]$.

(c) Since $L(f, P_n) = \sum_{i=n+1}^{2n} \frac{1}{i}$, we know that this sequence converges, because it is increasing and bounded above by 1. Use calculus to find the value of this convergent sequence. See Exercise 6.7.

10. Prove that if $h(x) \geq 0$ for all $x \in [a, b]$, then $\underline{\int_a^b} h \geq 0$ and $\overline{\int_a^b} h \geq 0$.

11. In Lemma 24.1, prove that $U(f, P^*) \leq U(f, P)$.

12. Convince yourself of the difficulty in applying Definition 4 to as simple a function as $f(x) = x$ on the interval $[1, 2]$.

13. Prove that if the function f is integrable on $[a, b)$ with a jump discontinuity at b, then f is integrable on $[a, b]$.

14. Let $f(x)$ be defined on $(a, b]$ and $\lim_{x \to a+} f(x) = +\infty$. Prove that $U(f, P) = +\infty$ for any partition P of $[a, b]$, and hence $\overline{\int_a^b} f = +\infty$. This is why we only worked with bounded functions in this section, because unbounded functions can never be integrable according to Definition 7.

15. Show that $U(f, P_n) = \frac{15}{2} + \frac{9}{2n}$ and $\inf_{P_n} U(f, P_n) = \frac{15}{2}$. Similarly show that $L(f, P_n) = \frac{15}{2} - \frac{9}{2n}$ and $\sup_{P_n} L(f, P_n) = \frac{15}{2}$.
Then tell why $\underline{\int_1^4} x \, dx = \overline{\int_1^4} x \, dx = \frac{15}{2}$.

SECTION 25 PROPERTIES OF THE RIEMANN INTEGRAL

We will prove some important theorems about integrals in Sections 26 and 27. Some of these proofs need the basic properties of the definite integral, which we state and prove in this section. All these results are called lemmas, instead of theorems, because they are mainly used to help prove theorems. These results are proved by the use of Definition 24.7 and Theorem 24.4. Whenever we use the term "integrable," we will mean Riemann integrable in the sense of Definition 24.7.

LEMMA 25.1   If f is integrable on $[a, b]$, then kf is integrable on $[a, b]$ for any constant k, and $\int_a^b kf = k \int_a^b f$.

Proof.   Let $\epsilon > 0$ be given. In order to prove that kf is integrable on $[a, b]$, we must find a partition P of $[a, b]$ so that $U(kf, P) - L(kf, P) < \epsilon$.
Case 1. $k > 0$

Since f is integrable on $[a, b]$, there is a partition P_ϵ of $[a, b]$ so that $U(f, P_\epsilon) - L(f, P_\epsilon) < \frac{\epsilon}{k}$. But

$$U(kf, P_\epsilon) - L(kf, P_\epsilon) = k\, U(f, P_\epsilon) - k\, L(f, P_\epsilon)$$

$$= k\left(U(f, P_\epsilon) - L(f, P_\epsilon)\right) < k \cdot \frac{\epsilon}{k} = \epsilon$$

We next need to show that $\int_a^b kf = k \int_a^b f$. This can be done by use of the following set of inequalities. You are asked to supply the reasons as an exercise.

$$k\, L(f, P_\epsilon) = L(kf, P_\epsilon) \leq \int_a^b kf \leq U(kf, P_\epsilon) = k\, U(f, P_\epsilon)$$

$$L(f, P_\epsilon) \leq \int_a^b f \leq U(f, P_\epsilon)$$

$$k\, L(f, P_\epsilon) \leq k \int_a^b f \leq k\, U(f, P_\epsilon) \qquad \text{since } k > 0$$

Therefore

$$\left|\left(\int_a^b kf\right) - \left(k \int_a^b f\right)\right| \leq U(kf, P_\epsilon) - L(kf, P_\epsilon) < \epsilon$$

and so $\int_a^b kf = \int_a^b f$ by Lemma 5.1.   ∎

Case 2. $k = 0$

The function $kf = 0$ is integrable by use of Example 24.1, and $\int_a^b 0f = 0 \int_a^b f = 0$. ∎

Case 3. $k < 0$

This case is left as an exercise.

Comment.   In the results known as the Product Rule, the Quotient Rule, and the Chain Rule in Chapter 5, if the functions f and g are given to be differentiable, then a single proof accomplishes two goals—it shows that the corresponding functions $(fg, \frac{f}{g}, \text{ and } f \circ g)$ are differentiable, and also gives the value for the derivative of these functions. The above proof of Lemma 25.1, as well as the proofs of most of the remaining results in this chapter, shows that the situation is different for integration theory. We must give one proof (usually using Theorem 24.4) to show that the appropriate function is integrable, and then a second proof to find the value of the integral in question. This pattern is repeated in the proofs of Lemmas 25.2, 25.3, and 25.5. Theorems 26.1, 26.2, 26.5, and 26.7 are statements that certain functions are integrable, but nothing can be said about the value of the integral in these cases. On the other hand, Theorem 27.2 is a statement about the value of the integral.

LEMMA 25.2   If f is integrable on $[a, b]$ and $[b, c]$, where $a < b < c$, then f is integrable on $[a, c]$ and $\int_a^c f = \int_a^b f + \int_b^c f$.

Proof.   In order to prove that f is integrable on $[a, c]$, for each $\epsilon > 0$, we must find a partition P_ϵ of $[a, c]$, so that $U(f, P_\epsilon) - L(f, P_\epsilon) < \epsilon$. We do have a partition P_1 of $[a, b]$, so that $U(f, P_1) - L(f, P_1) < \frac{\epsilon}{2}$, and a partition P_2 of $[b, c]$, so that $U(f, P_2) - L(f, P_2) < \frac{\epsilon}{2}$. The key to this proof is to see that $P_1 \cup P_2$ is a partition of $[a, c]$. If we let $P_\epsilon = P_1 \cup P_2$, we observe that

$$L(f, P_\epsilon) = L(f, P_1) + L(f, P_2)$$

and

$$U(f, P_\epsilon) = U(f, P_1) + U(f, P_2)$$

Therefore

$$U(f, P_\epsilon) - L(f, P_\epsilon) = \Big(U(f, P_1) + U(f, P_2)\Big) - \Big(L(f, P_1) + L(f, P_2)\Big)$$

$$= \Big(U(f, P_1) - L(f, P_1)\Big) - \Big(U(f, P_2) - L(f, P_2)\Big)$$

$$< \frac{\epsilon}{2} + \frac{\epsilon}{2} = \epsilon \quad \blacksquare$$

Thus the proof that f is integrable on $[a, c]$ is complete. However, we still must prove the second part of the conclusion about the value of the integrals. All the information we need is contained in the proof above, but we need to write it in a different form. In particular, since

$$L(f, P_\epsilon) \leq \int_a^c f \leq U(f, P_\epsilon)$$

$$L(f, P_1) \leq \int_a^b f \leq U(f, P_1)$$

$$L(f, P_2) \leq \int_b^c f \leq U(f, P_2)$$

it follows that

$$L(f, P_1) + L(f, P_2) \leq \int_a^b f + \int_b^c f \leq U(f, P_1) + U(f, P_2)$$

Recalling that

$$L(f, P_\epsilon) = L(f, P_1) + L(f, P_2) \leq \int_a^c f \leq U(f, P_1) + U(f, P_2) = U(f, P_\epsilon)$$

we finally have

$$\left| \left(\int_a^b f + \int_b^c f \right) - \left(\int_a^c f \right) \right| \leq U(f, P_\epsilon) - L(f, P_\epsilon) < \epsilon$$

This completes the proof that $\int_a^c f = \int_a^b f + \int_b^c f$. $\blacksquare$

Lemma 25.2 is an additivity result involving one function on two intervals, while Lemma 25.3 is an additivity result involving two functions on one interval. The proofs have a similar form, so you are encouraged to practice your skills by finishing the proof that is started below.

LEMMA 25.3   If f and g are both integrable on $[a, b]$, then $f + g$ is integrable on $[a, b]$, and $\int_a^b (f + g) = \int_a^b f + \int_a^b g$.

Proof.   Let $\epsilon > 0$ be given. Since f and g are both integrable on $[a, b]$, there are partitions P_1 and P_2 of $[a, b]$ so that

$$U(f, P_1) - L(f, P_1) < \frac{\epsilon}{2}$$

$$U(g, P_2) - L(g, P_2) < \frac{\epsilon}{2}$$

If P_ϵ is the refinement partition $P_1 \cup P_2$, then

$$U(f, P_\epsilon) - L(f, P_\epsilon) < \frac{\epsilon}{2}$$

$$U(g, P_\epsilon) - L(g, P_\epsilon) < \frac{\epsilon}{2}$$

The next step is to show that

$$U(f + g, P_\epsilon) \leq U(f, P_\epsilon) + U(g, P_\epsilon)$$

$$L(f + g, P_\epsilon) \geq L(f, P_\epsilon) + L(g, P_\epsilon)$$

which can be done by the use of Lemma 5.4. It is left as an exercise for you to finish this proof.   ■

There has been a limit theorem in each of the previous four chapters, and it would be natural to consider an analogous result here; that is, if two functions are integrable, will their sum, difference, product, quotient, and composition also be integrable, and can anything be said about the values for $\int f + g$, $\int f - g$, $\int fg$, $\int \frac{f}{g}$, and $\int g \circ f$, in terms of the values for $\int f$ and $\int g$? Lemmas 25.1 and 25.3 together imply that there is such a result for sums, scalar multiples, and differences. In the language of linear algebra, we could say that the collection of integrable functions on $[a, b]$ form a vector space. It is also true that the product is integrable, but that is a more complicated theorem to prove, and there is no formula for $\int_a^b fg$ in terms of $\int_a^b f$ and $\int_a^b g$. This question and the related one for quotients and composite functions will be discussed in the next section. In this section, we consider only those properties that can be proved quickly from the definition of the integral and the Riemann integrability criterion.

LEMMA 25.4   If f and g are integrable on $[a, b]$ and $f(x) \leq g(x)$ on $[a, b]$, then $\int_a^b f \leq \int_a^b g$.

Proof.   Let $h(x) = g(x) - f(x)$, and apply Exercise 24.10, along with Lemmas 25.1 and 25.2. ∎

LEMMA 25.5   If f is integrable on $[a, b]$, then $|f|$ is also integrable on $[a, b]$, and $\left| \int_a^b f \right| \leq \int_a^b |f|$.

Proof.   One way to prove that $|f|$ is integrable on $[a, b]$ is to use the result that

$$\sup_T |f| - \inf_T |f| \leq \sup_T f - \inf_T f$$

for any set T (see Exercise 5). From this, it follows that for any partition $P = \{x_i\}_{i=0}^n$ of $[a, b]$,

$$U(|f|, P) - L(|f|, P) = \sum_{i=1}^n \Big(M_i(|f|) - m_i(|f|) \Big) \Delta x_i$$

$$\leq \sum_{i=1}^n \Big(M_i(f) - m_i(f) \Big) \Delta x_i$$

$$= U(f, P) - L(f, P)$$

and therefore $|f|$ is integrable by Theorem 24.4.

To prove the second part of the theorem, we observe that $f(x) \leq |f(x)|$ implies that $\int_a^b f \leq \int_a^b |f|$ and $-f(x) \leq |f(x)|$ implies also that $\int_a^b (-f) \leq \int_a^b |f|$ by Lemma 25.4. Therefore $\left| \int_a^b f \right| \leq \int_a^b |f|$. Why? ∎

LEMMA 25.6   If f is integrable on $I = [a, b]$, then f is integrable on every subinterval J of I.

Proof.   Let $J = [c, d]$, where $a \leq c < d \leq b$, and let $\epsilon > 0$ be given. There is a partition P of $[a, b]$, so that $U(f, P) - L(f, P) < \epsilon$. Also $P' = P \cup \{c, d\}$ is another partition of $[a, b]$, for which

$$U(f, P') - L(f, P') \leq U(f, P) - L(f, P)$$

It is also true that P' can be written as $P_1 \cup P_2 \cup P_3$, where P_1 is a partition of $[a, c]$, P_2 is a partition of $J = [c, d]$, P_3 is a partition of $[d, b]$

and $U(f, P_2) - L(f, P_2) \leq U(f, P') - L(f, P')$. Why is this inequality true and why does it complete the proof? ∎

THEOREM 25.7 (Mean Value Theorem for Integrals)
If f is continuous on $[a, b]$, then there is some value $c \in [a, b]$ so that $\int_a^b f(x)\, dx = f(c) \cdot (b - a)$.

Comment. This theorem is often called the Theorem of the Mean. This result reminds us of the Greek goal to show that areas of curved regions are equal to areas of straight-edged regions, for it states that the area of a general region equals the area of a rectangle. However, this is an existence theorem and it does not give a method for finding the value for c. Cauchy justified this result by an averaging argument, and then used it to prove the Fundamental Theorem of the Calculus. Cauchy's proof for Theorem 25.7 was based on intuition, so we need to replace it by one based on previously established results.

Proof. By considering the geometry of areas and the Riemann sums used to define the definite integral, we have the inequality that

$$m(b - a) \leq \int_a^b f(x)\, dx \leq M(b - a)$$

where $m = \inf\{f(x) : x \in [a, b]\}$ and $M = \sup\{f(x) : x \in [a, b]\}$. From this it follows that

$$m \leq \frac{\int_a^b f(x)\, dx}{b - a} \leq M$$

Since f is continuous on $[a, b]$, by either the Extreme Value Theorem or the Intermediate Value Theorem, there is a value $c \in [a, b]$, so that

$$f(c) = \frac{\int_a^b f(x)\, dx}{b - a}$$

and the proof is complete. ∎

EXERCISES 25

1. Supply the reasons for the steps in the proof of Case 1 of Lemma 25.1.

2. Prove Case 3 of Lemma 25.1.

3. Finish the proof of Lemma 25.3. This lemma contains two assertions and each one needs a proof.

4. Prove Lemma 25.4, following the hint provided.

5. Prove the assertion stated in Lemma 25.5 that

$$\sup_T |f| \; - \; \inf_T |f| \; \le \; \sup_T f \; - \; \inf_T f$$

for any set T on which the function f is defined. This can best be done by considering cases.

6. Give the reason why $|\int_a^b f| \le \int_a^b |f|$ at the end of the proof for Lemma 25.5.

7. Give reasons for the questions asked at the end of the proof of Lemma 25.6.

8. Give an example of a function f so that f is not integrable on a given interval $[a, b]$ even though $|f|$ is integrable on $[a, b]$. This will show that the converse to Lemma 25.5 is not true.

9. Apply the Mean Value Theorem for Integrals to $f(x) = \sin x$ on $[0, \pi]$.

10. Prove that if f is continuous on $[a, b]$ except for jump discontinuities at a finite number of points, then f is integrable on $[a, b]$. Hint: use Lemma 25.2.

SECTION 26   TYPES OF INTEGRABLE FUNCTIONS

In previous chapters, we proved that sums, products, and quotients of continuous and differentiable functions were respectively continuous and differentiable. We also were able to decide the continuity and differentiability for familiar functions (e.g. $\sin x$ is continuous and differentiable for all x, while $\sqrt{x}$ is continuous for $x \geq 0$, and differentiable for $x > 0$). We now establish several general results about which functions are integrable. Of these, the following one is the most important. It was referred to in Chapter 4 as an important consequence of uniform continuity.

THEOREM 26.1   If f is continuous on the interval $[a, b]$, then f is integrable on $[a, b]$.

Proof.   Let $\epsilon > 0$ be given. According to Theorem 17.2, let $\delta > 0$ be such that if $x, u \in [a, b]$ and $|x - u| < \delta$, then $|f(x) - f(u)| < \frac{\epsilon}{b-a}$.

By the Archimedean axiom, choose n so that $\Delta x = \frac{b-a}{n} < \delta$, and let P_n be the partition of $[a, b]$ into n equal subintervals, each of length Δx.

For $x, u \in [x_{i-1}, x_i]$, $|x - u| \leq \Delta x < \delta$ and so $|f(x) - f(u)| < \frac{\epsilon}{b-a}$.

Since this is true for all $x, u \in [x_{i-1}, x_i]$, $M_i - m_i < \frac{\epsilon}{b-a}$ and

$$U(f, P_n) - L(f, P_n) = \sum_{i=1}^{n} (M_i - m_i) \Delta x$$

$$< \sum_{i=1}^{n} \frac{\epsilon}{b - a} \cdot \Delta x = \frac{\epsilon}{b - a} \cdot (b - a) = \epsilon$$

The function f is therefore integrable on $[a, b]$ by the Riemann integrability criterion in Theorem 24.4.   ∎

We next seek to decide what types of discontinuous functions will be integrable. We saw in Example 24.3 that the greatest integer function with a jump discontinuity is integrable. In Exercise 25.10, a function with a finite number of jump discontinuities is integrable. When we consider a function with infinitely many discontinuities, we recall Example 24.4 where a bounded function with uncountably many discontinuities was not integrable. However, a bounded function with a countable number of discontinuities will often be integrable (see the following example).

Example 1   Let $f(x) = \left[\frac{1}{x}\right]^{-1}$ on $(0,1]$ and $f(0) = 0$. f has a countably infinite number of jump discontinuities on $[0,1]$, but is still integrable
on this interval.

Solution.   To show this, we use Theorem 24.4 as follows. Let $\epsilon > 0$ be
given. Then f is integrable on $[0, \epsilon]$, because for any partition P of $[0, \epsilon]$,

$$U(f,P) - L(f,P) = \sum_{i=1}^{n}(M_i - m_i)\Delta x_i \leq \sum_{i=1}^{n}1\,\Delta x_i = \epsilon$$

The function f is also integrable on $[\epsilon, 1]$ because f has but a finite number
of jump discontinuities on this interval (see Exercise 25.10). Then f is
integrable on $[0,1]$ by Lemma 25.2.   □

Comment.   By a combination of previous results such as Lemma 19.1
and Theorem 26.1, we obtain the following interesting chain of set inequalities.

  Let $\mathcal{D}_n$ = collection of all functions defined on $[a, b]$ for which the first
            n derivatives exist.

  Let $\mathcal{D}_\infty$ = collection of all functions defined on $[a, b]$ for which deriva-
            tives of all orders exist.

  Let $\mathcal{C}$ = collection of all continuous functions defined on $[a, b]$.

  Let $\mathcal{I}$ = collection of all integrable functions on $[a, b]$.

  Let $\mathcal{B}$ = collection of all bounded functions on $[a, b]$.

These sets satisfy the following inequality chain.

$$\mathcal{D}_\infty \subset \cdots \subset \mathcal{D}_n \subset \cdots \subset \mathcal{D}_3 \subset \mathcal{D}_2 \subset \mathcal{D}_1 \subset \mathcal{C} \subset \mathcal{I} \subset \mathcal{B}$$

THEOREM 26.2   If f is monotone on the interval $[a, b]$, then f is
integrable on $[a, b]$.

Proof.   Assume first that f is nondecreasing on $[a, b]$. Let $\epsilon > 0$ be given,
and choose n so that $\big(f(b) - f(a)\big)(b - a) < \epsilon n$. Let $P_n = \{x_i\}_{i=0}^{n}$
partition $[a, b]$ into n equal parts. Then

$$U(f, P_n) - L(f, P_n) = \sum_{i=1}^{n}(M_i - m_i)\,\Delta x_i$$

$$= \sum_{i=1}^{n}\Big(f(x_i) - f(x_{i-1})\Big)\frac{b-a}{n}$$

$$= \frac{b-a}{n}\sum_{i=1}^{n}\Big(f(x_i) - f(x_{i-1})\Big)$$

$$= \frac{b-a}{n}\Big(f(b) - f(a)\Big) < \epsilon$$

Therefore f is integrable on $[a, b]$. ∎

We return now to the matter of a Limit Theorem for integrable functions. Because of Theorems 15.1 and 26.1, we know that if f and g are continuous on $[a, b]$, then the product fg, the quotient $\frac{f}{g}$ if $g \neq 0$ on $[a, b]$, and the composite $g \circ f$ will also be integrable functions because they are continuous. The question remains whether these results are true if f and g are integrable, but discontinuous, on $[a, b]$. Theorems 26.5, 26.7, and 26.8 contain an answer to this question. Several results about infimums and supremums are needed for the proofs of these theorems. A sample result is proved in Lemma 26.3 to provide a model for such proofs. You are asked to prove some similar results in the exercises.

LEMMA 26.3   Let T be any set on which the function g is defined. Then

$$\sup\{g(x) : x \in T\} - \inf\{g(x) : x \in T\} = \sup\{g(x) - g(y) : x, y \in T\}$$

Proof of $\geq$:   Choose any $x, y \in T$. Since $g(x) \leq \sup\{g(x) : x \in T\}$ and $g(y) \geq \inf\{g(x) : x \in T\}$, we have

$$g(x) - g(y) \leq \sup\{g(x) : x \in T\} - \inf\{g(x) : x \in T\}$$

Since this is true for all $x, y \in T$, by Lemma 5.3,

$$\sup\{g(x) - g(y) : x, y \in T\} \leq \sup\{g(x) : x \in T\} - \inf\{g(x) : x \in T\} \quad ∎$$

Proof of $\leq$:   Let $\epsilon > 0$ be given. By Lemma 5.2, there are $x, y \in T$ so that

$$g(x) > \sup_{T} g - \frac{\epsilon}{2} \quad \text{and} \quad g(y) < \inf_{T} g + \frac{\epsilon}{2}$$

From this, we obtain $g(x) - g(y) > \sup_T g - \inf_T g - \epsilon$, and then

$$\sup \{ g(x) - g(y) : x, y \in T \} \geq \sup_T g - \inf_T g - \epsilon$$

by Lemma 5.3. Finally

$$\sup \{ g(x) - g(y) : x, y \in T \} \geq \sup_T g - \inf_T g \quad \blacksquare$$

LEMMA 26.4   If f is integrable on $[a, b]$, then f^2 is also integrable on $[a, b]$.

Proof.   Let $|f(x)| \leq B$ for all $x \in [a, b]$. Then the key to this proof is the result in Exercise 6 that

$$\sup_T f^2 - \inf_T f^2 \leq 2B \left(\sup_T f - \inf_T f \right)$$

for any set T, obtained by the use of Lemma 26.3.

For $\epsilon > 0$, let $P = \{x_i\}_{i=0}^n$ be a partition of $[a, b]$ so that

$$U(f, P) - L(f, P) < \frac{\epsilon}{2B}$$

Apply Exercise 6 to sets of the form $[x_{i-1}, x_i]$, to obtain

$$U(f^2, P) - L(f^2, P) = \sum_{i=1}^n \left(M_i(f^2) - m_i(f^2) \right) \Delta x_i$$

$$\leq \sum_{i=1}^n 2B \left(M_i(f) - m_i(f) \right) \Delta x_i$$

$$= 2B \left(U(f, P) - L(f, P) \right) < \epsilon \quad \blacksquare$$

THEOREM 26.5   If f and g are integrable on $[a, b]$, then fg is also integrable on $[a, b]$.

Proof.   This proof is a simple consequence of Lemmas 25.2, 25.3, and 25.6. Since f and g are integrable on $[a, b]$, we know that $f + g$, $f - g$, $(f + g)^2$, and $(f - g)^2$ are also integrable. But so will fg be integrable, because $fg = \frac{1}{4} \left((f + g)^2 - (f - g)^2 \right)$. $\blacksquare$

You are again reminded that there is no "Product Rule" formula that relates the value of $\int_a^b fg$ to the values of $\int_a^b f$ and $\int_a^b g$. The integration by parts rule is the closest we come to such a formula.

The next example shows that it is not generally true that $\frac{f}{g}$ is integrable if f and g are integrable and $g \neq 0$ on $[a, b]$. However Lemma 26.6 contains a proof that this result is true if g is bounded away from 0.

Example 2   The function $f(x) = x$ if $x \neq 0$ and $f(0) = 1$ is integrable and non-zero on $[0, 1]$. However, $\frac{1}{f}$ is unbounded and therefore not integrable on $[0, 1]$.   □

LEMMA 26.6   If f is integrable on $[a, b]$ and there is some value $\alpha > 0$ so that $|f(x)| \geq \alpha$ for all $x \in [a, b]$, then $\frac{1}{f}$ is integrable on $[a, b]$.

Proof.   From the hypothesis, we can assume either that $f(x) \geq \alpha > 0$ or $f(x) \leq -\alpha < 0$ on $[a, b]$. It may be that $[a, b]$ needs to be partitioned into subintervals so that f has one of these two properties on each subinterval. For the proof below, we assume that $f(x) \leq -\alpha < 0$ on $[a, b]$. For each $\epsilon > 0$, there is a partition P of $[a, b]$ so that

$$U(f, P) - L(f, P) = \sum_{i=1}^{n} \Big(M_i(f) - m_i(f) \Big) \Delta x_i < \epsilon \alpha^2$$

Since $m_i(f) \leq M_i(f) \leq -\alpha < 0$, it follows that

$$\frac{1}{-m_i(f)} \leq \frac{1}{-M_i(f)} \leq \frac{1}{\alpha}$$

Then
$$U\left(\tfrac{1}{f}, P\right) - L\left(\tfrac{1}{f}, P\right) = \sum_{i=1}^{n} \left(M_i\left(\tfrac{1}{f}\right) - m_i\left(\tfrac{1}{f}\right) \right) \Delta x_i$$

$$= \sum_{i=1}^{n} \left(\frac{1}{m_i(f)} - \frac{1}{M_i(f)} \right) \Delta x_i \qquad \text{by Exercise 7}$$

$$= \sum_{i=1}^{n} \Big(M_i(f) - m_i(f) \Big) \Delta x_i \left(\frac{1}{-m_i(f)} \right) \left(\frac{1}{-M_i(f)} \right)$$

$$\leq \frac{1}{\alpha^2} \sum_{i=1}^{n} \Big(M_i(f) - m_i(f) \Big) \Delta x_i$$

$$= \frac{1}{\alpha^2} \Big(U(f, P) - L(f, P) \Big) < \epsilon \qquad ■$$

THEOREM 26.7   If f and g are integrable on $[a, b]$ and there is some constant $k > 0$, so that $|g(x)| \geq k$ on $[a, b]$, then $\frac{f}{g}$ is integrable on $[a, b]$.

Proof.   This is left as an exercise.

It is now possible to collect several previous results into a single limit theorem for integrable functions, which can be compared to Theorem 15.1 for continuous functions and Theorem 19.2 for differentiable functions.

THEOREM 26.8 (Limit Theorem for Integrable Functions)
If the functions f and g are integrable on $[a, b]$, then for all constants c_1 and c_2, the functions $c_1 f + c_2 g$ and fg are integrable on $[a, b]$. In addition, if $|g(x)| \geq k > 0$, then $\frac{f}{g}$ is integrable on $[a, b]$. As far as the value of the integrals are concerned, all we can prove is that

$$\int_a^b c_1 f + c_2 g = c_1 \int_a^b f + c_2 \int_a^b g$$

Proof.   See the results of Lemmas 25.1 and 25.3 along with Theorems 26.5 and 26.7.   ∎

Comment.   In writing a mathematics text, an author must sometimes decide whether to prove a general theorem that has several useful corollaries, or else to prove each corollary individually. Each approach has a significant advantage and disadvantage. The general theorem approach has the advantage that several results are obtained by a single proof, but the disadvantage is that this proof is usually abstract and non-intuitive, and gives little insight to the student. On the other hand, there is a duplication of effort to prove the results individually, but the student usually can follow the proofs more easily. For pedagogical purposes, the proof technique is probably learned better by such repetition.

Theorem 26.9 provides an illustration of such a general theorem, with a fairly long and detailed proof. If this theorem were proved first, then Lemmas 25.5, 26.4, and 26.6 become simple corollaries by letting $g(x)$ equal the functions $|x|$, x^2, and $\frac{1}{x}$ in that order. On the other hand, by proving the three results separately, we see the pattern of showing that $U(|f|, P) - L(|f|, P)$, $U(f^2, P) - L(f^2, P)$ and $U(\frac{1}{f}, P) - L(\frac{1}{f}, P)$ are in turn less than some constant times $U(f, P) - L(f, P)$.

It may be that some blending of these two approaches is desirable, with the proportion determined by the level of the course. A graduate level course would best concentrate on proving general theorems because it is the more powerful and elegant way. A first theory course past calculus should have results showing a repetition of method as a learning tool, and occasionally include a general theorem to prepare the student for the next course. In this setting, both approaches are mentioned, so you can make a comparison.

THEOREM 26.9   If f is integrable on the interval $I = [a, b]$ and g is continuous on the interval $[c, d]$ so that $f(I) \subset [c, d]$, then the composite function $g \circ f$ is integrable on I.

Proof.   See Theorem 6.2.12 in *Introduction to Real Analysis* by Robert Bartle and Donald Sherbert for a proof of this result.

EXERCISES 26

1. Prove Theorem 26.2 for the case where f is non-increasing on $[a, b]$.

2. Prove Theorem 26.7.

3. Use Theorem 26.9 to obtain a short proof of Lemma 26.4.

4. In what ways is the Limit Theorem for integrable functions (Theorem 26.8) different from the Limit Theorem for differentiable functions (Theorem 19.2)?

5. Let $|g(x)| \leq B$ for all $x \in T$. Show that

$$\sup\{f(x)g(x) : x \in T\} \leq B \sup\{f(x) : x \in T\}$$

6. If $|f(x)| \leq B$ for all $x \in T$, prove that

$$\sup_T f^2 - \inf_T f^2 \leq 2B\left(\sup_T f - \inf_T f\right)$$

Hint: Use Lemma 26.3 and Exercise 5.

7. Assume that $g(x) > 0$ for all $x \in T$. Prove that

$$\sup_T g = \frac{1}{\inf_T\left(\frac{1}{g}\right)} \quad \text{and} \quad \inf_T g = \frac{1}{\sup_T\left(\frac{1}{g}\right)}$$

This result is used in the proof of Lemma 26.6.

SECTION 27   INTEGRATION THEOREMS

It is with ill-concealed excitement that we come to this section and the proofs of the results that are known as the Fundamental Theorem of the Calculus. As previewed in Chapter 1, these theorems are statements about relationships among the concepts of continuity, differentiation, and integration. In Section 26, we proved that continuous functions are integrable. In this section, we prove results that can be intuitively expressed by saying that the integral is a continuous function and that differentiation and integration are inverse processes.

The key to the proof of the Fundamental Theorem is contained in the next lemma, where the Mean Value Theorem plays a crucial role. The proof of Theorem 27.2 itself will appear deceptively simple, and it will complete the theoretical chain of results begun with the assumption of the least upper bound axiom at the start of Chapter 2.

LEMMA 27.1   If the function f is defined and bounded on the closed interval $[a, b]$, and the function F is continuous on $[a, b]$, and $F' = f$ on the open interval (a, b), then every partition P of $[a, b]$ has a Riemann sum with a value equal to $F(b) - F(a)$.

Proof.   Let $P = \{x_i\}_{i=0}^n$ be a partition of $[a, b]$. We apply the Mean Value Theorem to F on each of the subintervals $[x_{i-1}, x_i]$. This is possible because F is continuous on each $[x_{i-1}, x_i]$, and differentiable on each (x_{i-1}, x_i).

So for each i from 1 to n, there is a value $c_i \in (x_{i-1}, x_i)$ so that

$$F'(c_i) = \frac{F(x_i) - F(x_{i-1})}{x_i - x_{i-1}}$$

Since $F'(c_i) = f(c_i)$ and since $\Delta x_i = x_i - x_{i-1}$, we can rewrite this in the form $F(x_i) - F(x_{i-1}) = f(c_i) \Delta x_i$. We then sum these values from 1 to n to obtain

$$\sum_{i=1}^n f(c_i) \Delta x_i = \sum_{i=1}^n F(x_i) - F(x_{i-1})$$

$$= F(x_n) - F(x_0) = F(b) - F(a)$$

and the proof of the lemma is complete.   ∎

THEOREM 27.2 (The Fundamental Theorem of Calculus)
If f is integrable on $[a, b]$, and F is continuous on $[a, b]$ with $F' = f$ on (a, b), then $\int_a^b f(x)\, dx = F(b) - F(a)$.

Proof. Let $\epsilon > 0$ be given. Since f is integrable on $[a, b]$, we have a partition P_ϵ of $[a, b]$, so that $U(f, P_\epsilon) - L(f, P_\epsilon) < \epsilon$. We recall that

$$L(f, P_\epsilon) \;\leq\; \int_a^b f(x)\, dx \;\leq\; U(f, P_\epsilon)$$

and
$$L(f, P_\epsilon) \;\leq\; \sum_{i=1}^{n} f(c_i)\Delta x_i \;\leq\; U(f, P_\epsilon)$$

for every Riemann sum determined by P_ϵ. By Lemma 27.1, there is a Riemann sum for f on P_ϵ that is equal to $F(b) - F(a)$, and therefore

$$L(f, P_\epsilon) \;\leq\; F(b) - F(a) \;\leq\; U(f, P_\epsilon)$$

Putting these statements together, we have

$$\left| \left(\int_a^b f(x)\, dx \right) - \Big(F(b) - F(a) \Big) \right| \;<\; \epsilon$$

and so $\int_a^b f(x)\, dx = F(b) - F(a)$.   ∎

In order to present the next theorems, we introduce the idea of a *function defined by an integral*. If $f(t)$ is integrable on an interval I containing a, then the function $F(x) = \int_a^x f(t)\, dt$ is defined for all x in I. We next prove that such a function is always continuous.

THEOREM 27.3 (The Integral Is a Continuous Function)
If $f(t)$ is integrable on an interval I containing a, then $F(x) = \int_a^x f(t)\, dt$ is continuous on I.

Proof. There is a value B so that $f(x) \leq B$ for all $t \in I$. Next choose a value $c \in I$ and assume that $x > c$. Then

$$| F(x) - F(c) | \;=\; \left| \int_a^x f(t)\, dt \;-\; \int_a^c f(t)\, dt \right|$$

$$= \left| \int_c^x f(t)\, dt \right|$$

$$\leq \int_c^x |f(t)|\, dt$$

$$\leq \int_c^x B\, dt = B(x - c)$$

You should cite an appropriate lemma from Section 25 as the reason for each of the above steps. A similar inequality follows if $x < c$, and so for any $\epsilon > 0$, if we let $\delta_\epsilon = \frac{\epsilon}{B}$, and choose x so that $|x - c| < \delta_\epsilon$, then

$$|F(x) - F(c)| \leq B\,|x - c| < B\frac{\epsilon}{B} = \epsilon$$

and $F(x)$ is continuous at c by use of the ϵ-δ definition of continuity. ∎

From previous work, we know that $f(t)$ can be integrable without being continuous. In such a case, $F(x) = \int_a^x f(t)\, dt$ is continuous, but may not be differentiable. An example of this was given in Example 2.5. The next theorem states that $F(x)$ will be differentiable at all points x so that $f(t)$ is continuous at x.

THEOREM 27.4   Let $f(t)$ be integrable on the interval I that contains the point a, and be continuous at $x_o \in I$. Then $F(x) = \int_a^x f(t)\, dt$ is differentiable at x_o, and $F'(x_o) = f(x_o)$.

Proof.   We begin with the limit needed to show that $F'(x_o)$ exists. By the definition of the derivative,

$$F'(x_o) = \lim_{x \to x_o} \frac{F(x) - F(x_o)}{x - x_o}$$

$$= \lim_{x \to x_o} \frac{\int_a^x f(t)\, dt - \int_a^{x_o} f(t)\, dt}{x - x_o}$$

$$= \lim_{x \to x_o} \frac{\int_{x_o}^x f(t)\, dt}{x - x_o}$$

$$= \lim_{x \to x_o} \frac{\int_{x_o}^x \Big(f(t) - f(x_o) + f(x_o) \Big)\, dt}{x - x_o}$$

$$= \lim_{x \to x_o} \left(\frac{\int_{x_o}^x \Big(f(t) - f(x_o) \Big)\, dt}{x - x_o} + \frac{\int_{x_o}^x f(x_o)\, dt}{x - x_o} \right)$$

$$= \lim_{x \to x_o} \frac{\int_{x_o}^x \Big(f(t) - f(x_o) \Big)\, dt}{x - x_o} + f(x_o)$$

By next showing that $\lim_{x \to x_o} \frac{\int_{x_o}^x \big(f(t) - f(x_o) \big)\, dt}{x - x_o} = 0$, we will have proved that $F'(x_o)$ exists and equals $f(x_o)$.

Let $\epsilon > 0$ be given. Since $f(t)$ is continuous at x_o, there is a $\delta > 0$ so that if $|t - x_o| < \delta$, then $|f(t) - f(x_o)| < \epsilon$. Choosing $x > x_o$, so that $|x - x_o| < \delta$, we have

$$\left| \frac{\int_{x_o}^x (f(t) - f(x_o))\, dt}{x - x_o} \right| \leq \frac{\int_{x_o}^x |f(t) - f(x_o)|\, dt}{|x - x_o|}$$

$$\leq \frac{\int_{x_o}^x \epsilon\, dt}{|x - x_o|} = \frac{\epsilon (x - x_o)}{x - x_o} = \epsilon$$

A similar result needs to be verified for $x < x_o$, and the proof of this theorem will then be complete. ■

Theorems 27.2 and 27.4 contain the two parts of the Fundamental Theorem statement that differentiation and integration are inverse processes. These are the formal statements, whereas Result 3 in Section 1 contained an intuitive expression of this idea. See Examples 5–7 in Section 2 for illustrations of the theorems contained in this section.

EXERCISES 27

1. In the proof of Theorem 27.3, give a reason for each of the steps.

2. Prove Theorem 27.3 for the case where $x < c$.

3. Verify Theorem 27.4 for the case where $x < x_o$.

4. Use Theorem 27.2 to evaluate the following integrals.

   (a) $\int_{-\pi}^{\pi} \sin nx \cos mx\, dx = 0$ for $n, m \in \mathbf{N}$.

   (b) $\int_{-\pi}^{\pi} \sin nx \sin mx\, dx = 0$ for $n \neq m$ in $\mathbf{N}$.

   (c) $\int_{-\pi}^{\pi} \sin^2 nx\, dx = \pi$

5. (a) If $f(t) = t - [t]$, find $F(x) = \int_0^x f(t)dt$ as was done in Example 2.5.

   (b) Illustrate Theorem 27.4 by applying it to $F(x)$ as defined in part (a).

6. Use integration by parts to prove that for any $x > 0$ and $k > 0$,

$$\int_0^k t^x e^{-t}\, dt = \frac{-k^x}{e^k} + x \int_0^k t^{x-1} e^{-t}\, dt$$

7. Use integration by parts twice to prove that

$$\int_{-\pi}^{\pi} f(x) \sin nx\, dx = (-1)^{n+1}\left(f(\pi) - f(-\pi) \right) - \frac{1}{n^2} \int_{-\pi}^{\pi} f''(x) \sin nx\, dx$$

assuming that f, f', and f'' are integrable functions.

8. Prove that if $f > 0$ on $[a, b]$, and $F(x) = \int_a^x f(t)\, dt$ is defined for $a \le x \le b$, then F is increasing on $[a, b]$.

9. Use L'Hospital's Rule and Theorem 27.4 to find

$$\lim_{x \to 0} \frac{\int_0^{x^2} \sin t^2\, dt}{x^2}$$

10. The function f defined on the interval $[-a, a]$ is called *even* if $f(-x) = f(x)$ for all $x \in [-a, a]$ and *odd* if $f(-x) = -f(x)$ for all $x \in [-a, a]$. Prove that $\int_{-a}^{a} f(x)\, dx = 2 \int_0^a f(x)\, dx$ if f is even, and $\int_{-a}^{a} f(x)\, dx = 0$ if f is odd.

SECTION 28 IMPROPER INTEGRALS

In the earlier sections of this chapter, the integral was defined only for functions that were bounded on a finite closed interval. Therefore functions such as $f(x) = \frac{1}{x^2}$ defined on the interval $[1, \infty)$ or $f(x) = \frac{1}{x^2}$ if $x \neq 0$ and $f(0) = 0$, defined on the interval $[0, 1]$, could not be integrable under the conditions of Definition 24.7.

Both functions determine a region in the plane for which we might try to define a number to represent the "area" of the unbounded region that they determine. The integrals $\int_1^\infty \frac{1}{x^2} \, dx$ and $\int_0^1 \frac{1}{x^2} \, dx$ are called *improper integrals*, and their values are defined as the limit of integrals that are well-defined Riemann integrals.

Definition 1   If f is integrable on intervals of the form $[a, t]$ for all $t > a$, then $\int_a^\infty f$ is an *improper integral of the first kind* and its value is defined to be $\int_a^\infty f = \lim_{t \to \infty} \int_a^t f$. Another form of an improper integral of the first kind is $\int_{-\infty}^a f$, where f is integrable on intervals of the form $[t, a]$, for all $t < a$. In this case, we define $\int_{-\infty}^a f = \lim_{t \to -\infty} \int_t^a f$.

Definition 2   If f is integrable on intervals of the form $[t, b]$ for $a < t \leq b$ and $\lim_{x \to a+} |f(x)| = \infty$, then $\int_a^b$ is an *improper integral of the second kind* and its value is defined to be $\int_a^b f = \lim_{t \to a+} \int_t^b f$. A similar improper integral is defined when f is integrable on intervals of the form $[a, t]$ for $a \leq t < b$ and $\lim_{x \to b-} |f(x)| = \infty$.

Definition 3   If the value of an improper integral of the first kind or the second kind is a real number, the integral is said to *converge* to that value. If the value is $+\infty$ (or $-\infty$), the improper integral is said to *diverge to* $+\infty$ (or $-\infty$). In all other cases, such as if the limit does not exist, the improper integral is simply said to *diverge*.

Example 1 $\int_1^\infty \frac{1}{x^2} \, dx$ is an improper integral of the first kind. It converges to 1 because

$$\int_1^\infty \frac{1}{x^2} \, dx = \lim_{t \to \infty} \int_1^t \frac{1}{x^2} \, dx = \lim_{t \to \infty} \left(\frac{-1}{t} + 1 \right) = 1 \quad \square$$

Example 2 $\int_0^1 \frac{1}{x^2}\, dx$ is an improper integral of the second kind. It diverges to $+\infty$ because

$$\int_0^1 \frac{1}{x^2}\, dx = \lim_{t \to 0+} \int_t^1 \frac{1}{x^2}\, dx = \lim_{t \to 0+} \left(-1 + \frac{1}{t} \right) = +\infty \quad \square$$

Example 3 $\int_{-\infty}^0 \sin x\, dx$ is improper of the first kind. It is divergent because

$$\lim_{t \to -\infty} \int_t^0 \sin x\, dx = \lim_{t \to -\infty} (-1 + \cos t)$$

does not exist because of oscillation. $\square$

Comment.   An integral is called *improper of the third kind* if it can be expressed as a sum of two or more improper integrals of the first or second kind. Such an integral will be said to converge only if all the improper integrals in its representation converge according to Definition 3. The value of a convergent improper integral of the third kind is the sum of the values of the convergent improper integrals in its representation.

Example 4 $\int_0^\infty \frac{1}{x^2}\, dx$ is improper of the third kind because we can use the representation

$$\int_0^\infty \frac{1}{x^2}\, dx = \int_0^1 \frac{1}{x^2}\, dx + \int_1^\infty \frac{1}{x^2}\, dx$$

It diverges because the improper integral of the second kind $\int_0^1 \frac{1}{x^2}\, dx$ diverges to $+\infty$. $\square$

In our work with infinite series in Chapter 8, the p-series $\sum_{n=1}^\infty \frac{1}{n^p}$ will be an important tool to use for establishing the convergence of other series by means of the Comparison Test. The convergence or divergence of the p-series, along with approximations for the value of the convergent series will be established in Section 31 by use of the following theorem. A related result was established in Example 2.6.

THEOREM 28.1   For $a > 0$, the improper integral $\int_a^\infty \frac{1}{x^p}\, dx$ converges to $\frac{a^{1-p}}{p-1}$ if $p > 1$, and diverges to $+\infty$ if $0 < p \le 1$.

Proof. If $p > 1$, then $\lim_{t \to \infty} \frac{1}{t^{p-1}} = 0$ and so

$$\int_a^\infty \frac{1}{x^p}\, dx \;=\; \lim_{t \to \infty} \int_a^t x^{-p}\, dx$$

$$=\; \lim_{t \to \infty} \frac{x^{-p+1}}{-p+1}\bigg|_a^t$$

$$=\; \lim_{t \to \infty} \frac{1}{1-p}\left(\frac{1}{t^{p-1}} - a^{1-p}\right)$$

$$=\; \frac{a^{1-p}}{p-1}$$

If $p = 1$, then

$$\int_a^\infty \frac{1}{x}\, dx \;=\; \lim_{t \to \infty} (\ln x)\bigg|_a^t \;=\; +\infty$$

If $p < 1$, then

$$\int_a^\infty \frac{1}{x^p}\, dx \;=\; \lim_{t \to \infty} \frac{x^{-p+1}}{-p+1}\bigg|_a^t \;=\; \lim_{t \to \infty} \frac{1}{1-p}(t^{1-p} - a^{1-p}) \;=\; +\infty \quad \blacksquare$$

In all the examples considered so far in this section, we decided whether an improper integral converged or diverged by finding an antiderivative and using Theorem 27.2. This method permitted us to also find the value of the convergent improper integrals. However, there are situations where it is either difficult or impossible to find an antiderivative for the given function. In such cases, we can use a "comparison test" to decide whether the given improper integral converges or diverges.

THEOREM 28.2   If $\int_a^\infty g(x)\, dx$ is a convergent improper integral of the first kind, f is integrable on $[a, t]$ for all $t \geq a$, and $0 \leq f(x) \leq g(x)$ for $x \geq a$, then $\int_a^\infty f(x)\, dx$ is also a convergent improper integral of the first kind.

Proof. If we define $G(t) = \int_a^t g(x)\, dx$, we know that $\lim_{t \to \infty} G(t)$ exists as a real number. Also $F(t) = \int_a^t f(x)\, dx$ is increasing on $[a, \infty)$ by Exercise 27.8, and bounded because $F(t) \leq G(t) \leq \lim_{t \to \infty} G(t)$ for

$t \geq a$. Therefore $\lim_{t\to\infty} F(t) = \int_a^\infty f(x)\,dx$ exists by Exercise 10.8, and so $\int_a^\infty f(x)\,dx$ is convergent. ∎

THEOREM 28.3   If $\int_a^b g$ is a convergent improper integral of the second kind with $\lim_{x\to a^+} g(x) = +\infty$, f is integrable on $[t, b]$ for $a < t < b$, $\lim_{x\to a^+} f(x) = +\infty$, and $0 \leq f(x) \leq g(x)$ for $a < x \leq b$, then $\int_a^b f(x)\,dx$ is also a convergent improper integral of the second kind.

Proof.   This is left as an exercise.

We now consider three examples that will permit us to define the Gamma function.

Example 5   If $x \geq 1$, then $f(t) = t^{x-1}e^{-t}$ is integrable on the interval $[0, 1]$, because $f(t)$ is continuous on $[0, 1]$. Hence $\int_0^1 t^{x-1}e^{-t}\,dt$ exists as a proper integral for all $x \geq 1$. □

Example 6   If $0 < x < 1$, then $\int_0^1 t^{x-1}e^{-t}\,dt$ is improper convergent of the second kind.

Solution.   For $t \geq 0$, $0 < e^{-t} \leq 1$, and so $0 < t^{x-1}e^{-t} \leq t^{x-1}$. Using Exercise 4, $\int_0^1 t^{x-1}\,dt$ converges as long as $0 < 1 - x < 1$, which is for values of x for which $0 < x < 1$, and therefore $\int_0^1 t^{x-1}e^{-t}\,dt$ converges by Theorem 28.3. □

Example 7   If $x > 0$, then $\int_1^\infty t^{x-1}e^{-t}\,dt$ is improper convergent of the first kind.

Solution.   Since $\lim_{t\to\infty} \frac{t^{x+1}}{e^t} = 0$ by Exercise 22.4, there is some value $c > 1$ so that $\frac{t^{x+1}}{e^t} \leq 1$ for all $t \geq c$.

From this it follows that $e^{-t}t^{x-1} \leq t^{-2}$ for $t \geq c$.

Since we know that $\int_c^\infty t^{-2}\,dt$ converges, $\int_c^\infty e^{-t}t^{x-1}\,dt$ also converges by Theorem 28.2. Since $e^{-t}t^{x-1}$ is integrable on $[1, c]$, it follows that

$$\int_1^\infty t^{x-1}e^{-t}\,dt = \int_1^c t^{x-1}e^{-t}\,dt + \int_c^\infty t^{x-1}e^{-t}\,dt$$

is convergent of the first kind. □

From Examples 5–7, it follows that the improper integral $\int_0^\infty t^{x-1}e^{-t}\,dt$ converges to a well-defined real number for any $x > 0$, and therefore the following function can be defined by this expression.

Definition 4 $\Gamma(x) = \int_0^\infty t^{x-1}e^{-t}\,dt$ is a well-defined function for all $x > 0$ and is called the *Gamma function*.

LEMMA 28.4   If $x > 0$, then $\Gamma(x+1) = x\,\Gamma(x)$.

Proof.

$$\Gamma(x+1) = \int_0^\infty t^x e^{-t}\,dt$$

$$= \lim_{k\to\infty} \int_0^k t^x e^{-t}\,dt$$

$$= \lim_{k\to\infty} \frac{-t^x}{e^t}\bigg|_0^k + \lim_{k\to\infty} \int_0^k xt^{x-1}e^{-t}\,dt$$

$$= \lim_{k\to\infty} \left(\frac{-k^x}{e^k}\right) + x\int_0^\infty t^{x-1}e^{-t}\,dt$$

$$= x\,\Gamma(x)$$

by the use of Exercises 22.4 and 27.6.   ∎

LEMMA 28.5 $\Gamma(n) = (n-1)!$ for $n \in \mathbf{N}$.

Proof. This result is proved by mathematical induction.

For $n = 1$,

$$\Gamma(1) = \int_0^\infty e^{-t}\,dt = \lim_{k\to\infty} -e^{-t}\bigg|_0^k = 1 = 0!$$

Assuming that $\Gamma(n) = (n-1)!$ is true, we have by Lemma 28.4,

$$\Gamma(n+1) = n\,\Gamma(n)$$

$$= n\,(n-1)! = n!\quad\blacksquare$$

Example 8   Show that $\Gamma(\frac{1}{2}) = \int_0^\infty e^{-x^2}\, dx = \sqrt{\pi} \approx 1.77245$.
See also Example 2.9.

Solution.

$$\Gamma\left(\frac{1}{2}\right) = \int_0^\infty t^{-\frac{1}{2}} e^{-t}\, dt$$

$$= \int_0^\infty 2e^{-x^2}\, dx \qquad \text{by using the substitution } t = x^2$$

$$= 2\,\frac{\sqrt{\pi}}{2} \qquad \text{by Example 2.9}$$

$$= \sqrt{\pi} \approx 1.77245 \quad \square$$

EXERCISES 28

1. Test the following improper integrals for convergence or divergence. Find the value of the convergent integrals.

   (a) $\int_1^\infty \frac{\ln x}{x^2}\, dx$

   (b) $\int_1^\infty \frac{1}{\sqrt{x}}\, dx$

   (c) $\int_0^1 \frac{1}{\sqrt{x}}\, dx$

2. Test the following improper integrals for convergence or divergence by a comparison test.

   (a) $\int_0^\infty e^{-x^2}\, dx$

   (b) $\int_0^1 \frac{1}{\sqrt{x(x-1)^2}}\, dx$

3. (a) If $\int_0^2 f(x)\, dx$ is improper of the third kind, what can you conclude about $f(x)$ on $[a, b]$?

   (b) If $\int_1^\infty f(x)\, dx$ is improper of the third kind, what can you conclude about $f(x)$?

4. Prove that $\int_a^b \frac{1}{(x-a)^p}\, dx$ converges if $0 < p < 1$ and diverges if $p \geq 1$.

5. If $0 \leq f(x) \leq g(x)$ on the interval $(a, b]$, f is integrable on $[t, b]$ for $a < t < b$, and $\int_a^b g(x)\, dx$ is a convergent improper integral, prove that $\int_a^b f(x)\, dx$ is either a proper integral or a convergent improper integral.

6. Use Theorem 28.4 and Example 8 to find values for $\Gamma(\frac{3}{2})$ and $\Gamma(\frac{5}{2})$.

7. Prove that $\int_0^\infty x^2 e^{-nx}\, dx = \frac{2}{n^3}$.

8. Prove that $\int_{-\infty}^\infty \frac{1}{\sqrt{2\pi}} e^{-\frac{x^2}{2}}\, dx = 1$.

9. Prove that $\int_0^\infty e^{-x^3}\, dx = \frac{1}{3}\Gamma(\frac{1}{3})$.

10. Prove that $\int_0^1 (\ln\frac{1}{u})^{x-1}\, du = \Gamma(x)$, by using the substitution $t = \ln\frac{1}{u}$.

11. Use the substitution $u = \frac{1}{1+e^t}$ in order to prove that

$$\int_{-\infty}^\infty \frac{e^{at}}{1+e^t}\, dt = \Gamma(a)\,\Gamma(1-a)$$

12. Prove Theorem 28.3.

7

Sequences of Functions

SECTION 29 POINTWISE AND UNIFORM CONVERGENCE

The next three chapters will deal with the topic of convergence, which can be studied in the four different settings shown in Figure 29.1.

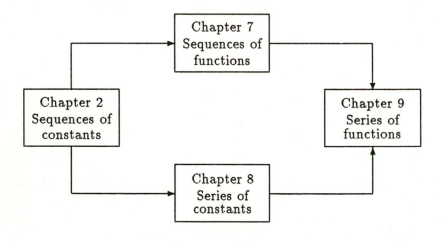

Figure 29.1

The above chart indicates that Chapters 7 and 8 can be covered in either order. However the theorems in Section 30 in Chapter 7 require some results from Chapters 4–6, whereas Chapter 8 could come directly after Chapter 2.

To compare these four settings by using topics already studied, we can identify $\{(1 + \frac{1}{n})^n\}$ as as example of a sequence of constants, while the p-series $\sum_1^\infty \frac{1}{n^2}$ is an example of a series of constants. Taylor's Theorem provides examples both of a sequence of functions and a series of functions.

A sequence of functions is given by $\{R_n(x)\}$, where the remainder term is

$$R_n(x) \;=\; \frac{f^{(n+1)}(c)}{(n+1)!}(x-a)^{n+1}$$

and c is some value between 0 and x. We might use $c(x, n)$ to emphasize that c depends both on x and n. If $\lim_{n\to\infty} R_n(x) = 0$, then we obtain an infinite Taylor polynomial $\sum_{n=0}^{\infty} \frac{f^{(n)}(a)}{n!}(x-a)^n$, which is an infinite series of functions that represents the function f. For example, if $f(x) = e^{2x}$ and $a = 0$, then $\sum_{n=0}^{\infty} \frac{2^n}{n!}x^n$ is the infinite Taylor polynomial (also called the Maclaurin series) for e^x, while the remainder sequence term is

$$R_n(x) \;=\; \frac{(2x)^{n+1}e^{2c}}{(n+1)!}$$

Definition 1   The sequence of functions $\{f_n(x)\}$ *converges pointwise to the limit function* $f(x)$ *on the set* S *if for each value* $x_o \in S$, *the sequence of constants* $\{f_n(x_o)\}$ *converges to* $\{f(x_o)\}$ *in the sense of Definition 4.2.* This means that for each $x_o \in S$ and for each $\epsilon > 0$, there is some $N(x_o, \epsilon)$ so that $|f_n(x_o) - f(x_o)| < \epsilon$ for all $n > N(x_o, \epsilon)$.

This kind of convergence is called *pointwise convergence* because the value for N depends on the point $x_o \in S$, as well as on ϵ, in contrast to the concept of uniform convergence defined below, where the value for N is independent of the choice of points in S.

Definition 2   The sequence of functions $\{f_n(x)\}$ *converges uniformly to the limit function* $f(x)$ *on the set* S *if for every* $\epsilon > 0$, *there is some* $N(\epsilon)$ so that $|f_n(x) - f(x)| < \epsilon$ for all $n > N(\epsilon)$ and for all $x \in S$.

We can also use our technique of negating a definition to obtain a description of when a sequence of functions will not converge uniformly on a set. This definition is somewhat complicated to apply, so you will be relieved to see the corollaries in Section 30, which often provide an easier way to prove that the convergence is not uniform.

Definition 3   The sequence of functions $\{f_n(x)\}$ *does not converge uni-formly to f on the set S* if there is some $\epsilon_o > 0$, so that for each $n \in \mathbf{N}$, there is some $n_o > n$, and some $x_n \in S$, so that $|f_{n_o}(x_n) - f(x_n)| \geq \epsilon_o$. Equivalently, there is a sequence $\{n_j\}$ of positive integers for which $n_1 < n_2 < \cdots < n_j \cdots$, and a sequence $\{x_j\}$ of points in S, so that $|f_{n_j}(x_j) - f(x_j)| \geq \epsilon_o$.

LEMMA 29.1   Let the sequence $\{f_n(x)\}$ converge pointwise to $f(x)$ on the set S. Choose $x_o \in S$ and a sequence $\{x_n\}$ so that $x_n \in S$ for all n. If $\lim_{n\to\infty} x_n = x_o$ and $\lim_{n\to\infty} f_n(x_n) \neq f(x_o)$, then $\{f_n(x)\}$ does not converge uniformly on the set S.

Proof.   This is left as an exercise.

Example 1   Let $f_n(x) = x^n$. Find the limit function $f = \lim_{n\to\infty} f_n$, and the sets on which the convergence is uniform. See Figure 29.2.

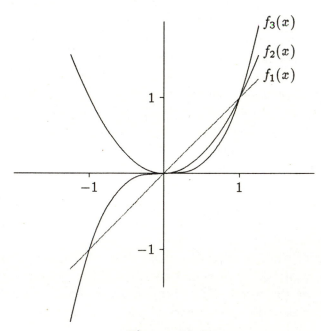

Figure 29.2

Solution.   The domain set S on which this sequence of functions is defined must be $(-1, +1]$ because $\{f_n(x)\}$ diverges for all other values of x. See Exercise 4.15, where we considered $\lim_{n\to\infty} c^n$. On this interval, we can see that the limit function will be

$$f(x) = \begin{cases} 0 & \text{if } -1 < x < 1 \\ 1 & \text{if } \quad x = 1 \end{cases}$$

We observe that even though each f_n is continuous on S, f is not continuous on S. Theorem 30.1 proves this cannot happen if the convergence is uniform.

We next use Definition 2 to prove that $\{x_n\}$ converges uniformly to 0 on sets of the form $[-a, +a]$, where $a < 1$. Since $\lim_{n\to\infty} a^n = 0$, for every $\epsilon > 0$, there is some N (i.e. $N = \left[\frac{\ln \epsilon}{\ln a}\right]$), so that $|a^n - 0| < \epsilon$ for all $n > N$. So if $x \in [-a, a]$ and $n \geq N$, then $|x^n - 0| \leq |x|^n \leq a^n < \epsilon$, for all $x \in [-a, a]$ and all $n \geq N$.

We finally apply Lemma 29.1 to prove that the convergence is not uniform on the set $(-1, +1]$. We concentrate our attention on the point 1, where the discontinuity of f occurs. If we let $x_n = 1 - \frac{1}{n}$, then

$$\lim_{n\to\infty} f_n(x_n) \; = \; \lim_{n\to\infty} (1 - \frac{1}{n})^n \; = \; \frac{1}{e} \; \neq \; f(1) \quad \square$$

Example 2   Let $f_n(x) = \frac{nx}{1+nx}$. Find the limit function $f = \lim f_n$ and the sets on which the convergence is uniform.

Solution.   The function $f_n(x)$ has an infinite discontinuity at $x = -\frac{1}{n}$ and a horizontal asymptote of $y = 1$. If $x = 0$, then $f_n(0) = 0$ for all n, while if $x \neq 0$, then

$$\lim_{n\to\infty} f_n(x) \; = \; \lim_{n\to\infty} \frac{nx}{1 + nx} \; = \; \lim_{n\to\infty} \frac{1}{\frac{1}{nx} + 1} = 1$$

Therefore

$$f(x) = \begin{cases} 0 & \text{if } x = 0 \\ 1 & \text{if } x \neq 0 \end{cases}$$

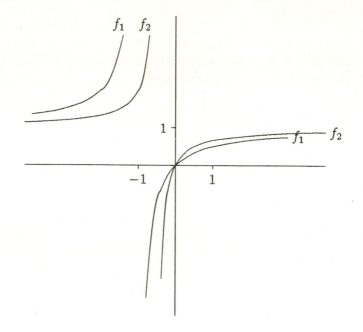

Figure 29.3

If $x_n = \frac{1}{n}$, we have $\lim_{n\to\infty} f_n(\frac{1}{n}) = \frac{1}{2} \neq 0 = f(0)$, and so the convergence is not uniform on any set containing 0. But we can prove that the convergence is uniform on sets of the form $S = [a, \infty)$, where $a > 0$, or of the form $(-\infty, b]$, where $b < 0$.

If $a > 0$ and $x \in [a, \infty)$, then

$$|f_n(x) - f(x)| = \left| \frac{nx}{1 + nx} - 1 \right| = \frac{1}{1 + nx} \leq \frac{1}{1 + na} < \epsilon$$

will hold if we choose

$$n > N = \left[\frac{\frac{1}{\epsilon} - 1}{a} \right]$$

For example, if $a = .01$ and $\epsilon = .0001$, then $N = 999,900$.

Notice that since each f_n is an increasing function, once we get the left endpoint value $f_n(a)$ into the ϵ-interval $(1 - \epsilon, 1 + \epsilon)$, then all values of $f_n(x)$ for $x > a$ will also get into this interval.

The proof of uniform convergence on $(-\infty, b]$, where $b < 0$ is left as an exercise. □

Example 3   Let $f_n(x) = xe^{-nx}$. Find the limit function $f = \lim f_n$ and the sets on which the convergence is uniform.

Solution.   Since $f_n'(x) = e^{-nx}(1 - nx) = 0$ when $x = \frac{1}{n}$, we know that the function f_n has a relative maximum at $(\frac{1}{n}, \frac{1}{en})$. Since $f_n(0) = 0$ and $\lim_{n \to \infty} f_n(x) = 0$, we observe that $|f_n(x)| \le \frac{1}{en}$ for all $x \ge 0$. For $x < 0$, $\lim_{n \to \infty} f_n(x) = -\infty$. Therefore $\{f_n\}$ converges pointwise to $f = 0$ on the set $S = [0, \infty)$.

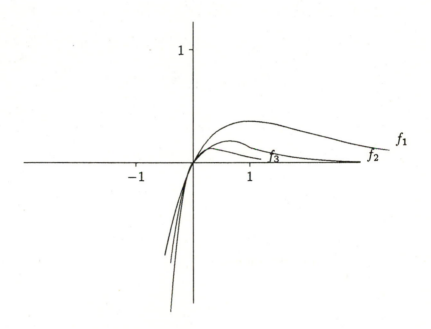

Figure 29.4

Let $\epsilon > 0$ be given and choose $N > \frac{1}{e\epsilon}$ so that $\frac{1}{eN} < \epsilon$. Then for all $x \ge 0$ and for all $n > N$, we have

$$|f_n(x) - 0| \le \frac{1}{en} < \frac{1}{eN} < \epsilon$$

and the convergence is uniform on $[0, \infty)$.   □

Comment.   The only method we have to show uniform convergence for the sequence $\{f_n\}$ is Definition 2, so we must first find the limit function f, and then show that $|f_n(x) - f(x)|$ can be made less than ϵ for all $x \in S$.

It is enough to find a sequence of constants $\{t_n\}$ so that $|f_n(x) - f(x)| \le t_n$ on S and $\lim t_n = 0$. In Example 1, we used $t_n = a^n$ on $S = [-a, +a]$. In Example 2, we used $t_n = \frac{1}{1+na}$ on $S = [a, \infty)$, and in Example 3, we used $t_n = \frac{1}{en}$ on $S = (-\infty, +\infty)$.

Example 4   Let $f_n(x) = ne^{-nx}$. Find the limit function $f = \lim f_n$, and the sets on which the convergence is uniform.

Solution.   We observe that for each n, f_n is a decreasing function with

$$\lim_{x \to -\infty} f_n(x) = +\infty, \quad f_n(0) = n, \quad \text{and} \quad \lim_{x \to \infty} f_n(x) = 0$$

If $x > 0$, we can use L'Hospital's Rule to show that

$$\lim_{n \to \infty} f_n(x) = \lim_{n \to \infty} \frac{n}{e^{nx}} = \lim_{n \to \infty} \frac{1}{xe^{nx}} = 0.$$

Therefore the limit function is $f = 0$ on the set $S = (0, \infty)$.

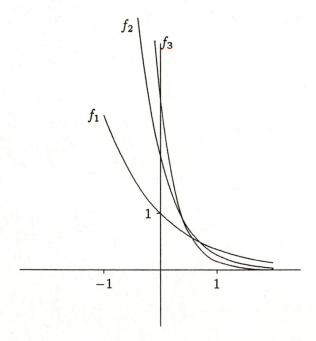

Figure 29.5

If the convergence to 0 of the sequence $\{f_n\}$ were uniform on $(0, \infty)$, then $|f_n(x) - 0| < \epsilon$ would have to hold for all $x > 0$, and all $n > N(\epsilon)$. But this cannot happen because $f_n(\frac{1}{n}) = \frac{n}{e}$.

However, the convergence is uniform on sets of the form $[a, \infty)$ for $a > 0$, because

$$|f_n(x) - f(x)| = |ne^{-nx} - 0| \le ne^{-na} = t_n$$

and the bounding sequence $\{t_n\}$ has the property that

$$\lim_{n \to \infty} t_n = \lim_{n \to \infty} ne^{-na} = 0 \quad \square$$

EXERCISES 29

Make a careful study of the following sequences of functions in problems 1–8 by the five steps given below.

(a) Make a graph of the functions f_1, f_2, and f_3.

(b) Find the set S on which $\{f_n\}$ converges pointwise.

(c) Find the limit function $f = \lim_{n \to \infty} f_n$.

(d) Based on your work done in parts (a), (b) and (c), make a guess as to the sets on which the convergence will be uniform and/or not uniform.

(e) Use Definitions 2 and 3 and Lemma 29.1 to prove your guesses in part (d).

1. $f_n(x) = \frac{x^n}{n}$

2. $f_n(x) = (\sin x)^n$

3. $f_n(x) = e^{-nx}$

4. $f_n(x) = \frac{x^n}{1 + x^n}$

5. $f_n(x) = \arctan nx$

6. $f_n(x) = nxe^{-nx}$

7. $f_n(x) = xe^{-nx^2}$

8. $f_n(x) = \begin{cases} n^2 x & \text{if } 0 \le x \le \frac{1}{n} \\ -n^2(x - \frac{2}{n}) & \text{if } \frac{1}{n} < x < \frac{2}{n} \\ 0 & \text{if } \frac{2}{n} \le x \end{cases}$

9. Prove Lemma 29.1.

10. In Example 2, prove that $\{f_n\}$ converges uniformly on $(-\infty, b]$ for $b < 0$.

SECTION 30   INTERCHANGE OF LIMIT THEOREMS

In Chapter 6, we proved several theorems that expressed the relationship between the ideas of continuity, differentiation and integration. In this section, we prove three important theorems that answer the question—if each f_n is continuous, integrable, or differentiable, will this property carry over to $\lim_{n \to \infty} f_n$? The concept of uniform convergence plays a significant role in each answer; that is why we laid the foundation in the last section in order to become comfortable with this idea.

THEOREM 30.1 (Uniform Convergence and Continuity)
If each function f_n is continuous on the set S, and if the sequence $\{f_n(x)\}$ converges uniformly to f on S, then f is continuous on S.

Proof. This proof is a classic analysis proof, in that we use the fact that each of three quantities can be made less than $\frac{\epsilon}{3}$. It is important to get the steps in the proper order. Let $\epsilon > 0$ and $x_o \in S$ be given. By uniform convergence, there is some N so that $|f_n(x) - f(x)| < \frac{\epsilon}{3}$ for all $n \geq N$, and for all $x \in S$. Since the specific function $f_N(x)$ is continuous at x_o, there is some $\delta(\epsilon, x_o) > 0$ so that $|f_N(x) - f_N(x_o)| < \frac{\epsilon}{3}$ for all x for which $|x - x_o| < \delta(\epsilon, x_o)$. But for these values of x, we have

$$|f(x) - f(x_o)| \;=\; |f(x) - f_N(x) + f_N(x) - f_N(x_o) + f_N(x_o) - f(x_o)|$$

$$\leq \; |f(x) - f_N(x)| + |f_N(x) - f_N(x_o)| + |f_N(x_o) - f(x_o)|$$

$$< \; \frac{\epsilon}{3} + \frac{\epsilon}{3} + \frac{\epsilon}{3} = \epsilon$$

and therefore f is continuous at x_o by the ϵ–δ definition of continuity. ∎

Historical comment. There is an interesting bit of history connected with the proof of Theorem 30.1. Throughout the eighteenth century, there was no distinction between pointwise convergence and uniform convergence. It was believed that any convergent sequence of continuous functions must converge to a continuous function. In the early 1800s, however, Fourier provided an example of a trigonometric series of continuous functions for which the limit function had jump discontinuities. In 1821,

Cauchy gave a proof for the result in Theorem 30.1, but he did not explicitly assume any concept of uniform convergence. Therefore his proof
must have been incorrect in light of the Fourier example. Much has been
written about this supposed error in Cauchy's work. The mathematical
community, including some of the best mathematical minds such as Abel
and Dirichlet, was aware of this apparent discrepancy between Cauchy's
proof and Fourier's example, and spent much time in a discussion of possible answers. Finally in 1847, P. L. Seidel discovered the hidden assumption
that Cauchy had used in his earlier proof. When this assumption, which
is equivalent to uniform convergence, is stated explicitly, then Cauchy's
proof is fully correct. The particular Fourier series is no longer a counterexample because it does not converge uniformly everywhere. The nature of this hidden assumption is akin to the nature of the error in the
incorrect proof we suggested earlier for Theorem 16.1. In that case, if the
domain of definition for the function is not a closed set, then the value
$B = \sup\{B_1, \ldots, B_n, \ldots\}$ could be infinite. In Cauchy's proof of Theorem 30.1, he assumed that there would be a largest value N in the set
$\{N(\epsilon, x) : x \in S\}$, where $N(\epsilon, x)$ is defined as a positive integer with
the property that if $n > N(\epsilon, x)$, then $|f_n(x) - f(x)| < \epsilon$. But if the
convergence is not uniform, then $\sup\{N(\epsilon, x) : x \in S\}$ could be infinite.

COROLLARY 30.2   Let f_n be continuous on the set S for all n, and let
the sequence $\{f_n(x)\}$ converge pointwise to f on S. If f is not continuous
on S, then $\{f_n(x)\}$ does not converge uniformly on S.

Proof.   This is the contrapositive form of Theorem 30.1.   ∎

This corollary provides an alternate means to show in Example 29.1
that the sequence $\{x^n\}$ does not converge uniformly on the set $(-1, +1]$,
and to show in Example 29.2 that the sequence $\{\frac{nx}{1+nx}\}$ does not converge
uniformly on the interval $[0, \infty)$. Neither the theorem nor the corollary
can be used to prove where the convergence is uniform however.

Example 1   Use Taylor's Theorem and Theorem 30.1 to show that

$$e = \sum_{n=0}^{\infty} \frac{1}{n!}$$

Solution. By Taylor's Theorem, we know that $e^x = f_n(x) + R_n(x)$ where

$$f_n(x) = 1 + x + \frac{x^2}{2!} + \cdots + \frac{x^n}{n!}$$

and for some value $c_n \in (0, x)$,

$$R_n(x) = \frac{e^{c_n}}{(n+1)!} x^{n+1}$$

Since e^x is continuous at $x = 1$, we have $e^1 = \lim_{x \to 1} e^x$.

But $\lim_{x \to 1} e^x = \lim_{x \to 1} \lim_{n \to \infty} \left(f_n(x) + R_n(x) \right)$

$$= \lim_{n \to \infty} \lim_{x \to 1} \left(f_n(x) + R_n(x) \right)$$

$$= \lim_{n \to \infty} \left(f_n(1) + R_n(1) \right)$$

$$= \lim_{n \to \infty} \left(1 + 1 + \frac{1}{2!} + \frac{1}{3!} + \cdots + \frac{1}{n!} \right) + \lim_{n \to \infty} \frac{e^{c_n}}{(n+1)!}$$

$$= \sum_{n=0}^{\infty} \frac{1}{n!} \qquad \square$$

The following comments illustrate what we mean by saying that Theorem 30.1 is a statement about the interchange of the order of limits, namely that

$$\lim_{x \to x_o} \lim_{n \to \infty} f_n(x) = \lim_{n \to \infty} \lim_{x \to x_o} f_n(x)$$

This can be observed by first noting that

$$\lim_{n \to \infty} \lim_{x \to x_o} f_n(x) = \lim_{n \to \infty} f_n(x_o) \qquad \text{since } f_n \text{ is continuous at } x_o$$

$$= f(x_o) \qquad \text{by the definition of } f$$

Also, $\lim_{x \to x_o} \lim_{n \to \infty} f_n(x) = \lim_{x \to x_o} f(x) \qquad \text{by the definition of } f$

$$= f(x_o) \qquad \text{since } f \text{ is continuous at } x_o$$

Every calculus student thinks of Theorem 27.2 whenever mention is made of the Fundamental Theorem of Calculus. If we broaden our idea of a fundamental theorem of analysis to include results about an interchange of limit process, then the three theorems in this section also qualify as fundamental theorems. These three theorems are restated in series format in Section 34, and those will also be fundamental theorems.

THEOREM 30.3 (Uniform Convergence and Integration)
If the sequence $\{f_n(x)\}$ converges uniformly to $f(x)$ on $[a, b]$, and each $f_n(x)$ is integrable on $[a, b]$, then $f(x)$ is integrable on $[a, b]$ and

$$\int_a^b f(x)\,dx = \int_a^b \lim_{n\to\infty} f_n(x)\,dx = \lim_{n\to\infty} \int_a^b f_n(x)\,dx$$

Comment. If each f_n is continuous, then f is continuous on $[a, b]$ by Theorem 30.1, and therefore integrable on $[a, b]$ by Theorem 26.1. If the f_n are integrable, but not necessarily continuous, then the proof below is needed to show that f is integrable.

Proof. Let $\epsilon > 0$ be given. To apply the Riemann integrability criterion, we need to find a partition P_ϵ of $[a, b]$ so that $U(f, P_\epsilon) - L(f, P_\epsilon) < \epsilon$. We first observe that because of uniform convergence, there is a positive integer N so that

$$|f_n(x) - f(x)| < \frac{\epsilon}{3(b-a)}$$

for all $n \geq N$ and for all $x \in [a, b]$. Since f_N is integrable on $[a, b]$, there is a partition $P_\epsilon = \{x_i\}_{i=0}^n$ of $[a, b]$, so that $U(f_N, P_\epsilon) - L(f_N, P_\epsilon) < \frac{\epsilon}{3}$.

It next follows that

$$\frac{-\epsilon}{3(b-a)} \leq m_i(f - f_N) \leq M_i(f - f_N) \leq \frac{\epsilon}{3(b-a)}$$

where the infimum values m_i and the supremum values M_i are taken over each set of the form $[x_{i-1}, x_i]$, for $i = 1, 2, \ldots, n$.

So using Lemma 5.4,

$$M_i(f) = M_i(f - f_N + f_N)$$

$$\leq M_i(f - f_N) + M_i(f_N)$$

$$\leq \frac{\epsilon}{3(b-a)} + M_i(f_N)$$

from which it follows that

$$U(f, P_\epsilon) = \sum_{i=1}^{n} M_i(f)\, \Delta x_i$$

$$\leq \sum_{i=1}^{n} \left(\frac{\epsilon}{3(b-a)} + M_i(f_N) \right) \Delta x_i$$

$$= \frac{\epsilon}{3(b-a)} \cdot (b-a) + U(f_N, P_\epsilon)$$

$$= \frac{\epsilon}{3} + U(f_N, P_\epsilon)$$

In a similar way, we can show that $-L(f, P_\epsilon) \leq -L(f_N, P_\epsilon) + \frac{\epsilon}{3}$.

Putting this all together, we conclude that

$$U(f, P_\epsilon) - L(f, P_\epsilon) \leq \left(U(f_N, P_\epsilon) + \frac{\epsilon}{3} \right) + \left(-L(f_N, P_\epsilon) + \frac{\epsilon}{3} \right)$$

$$= U(f_N, P_\epsilon) - L(f_N, P_\epsilon) + \frac{2\epsilon}{3}$$

$$< \frac{\epsilon}{3} + \frac{2\epsilon}{3} = \epsilon$$

and the proof is complete for the first part of the theorem. ■

The second part of the theorem is easily proved by showing that the sequence of constants $\{ \int_a^b f_n(x)\, dx \}$ converges to the real number $\int_a^b f(x)\, dx$ by using Definition 4.2. Because of the uniform convergence, for a given $\epsilon > 0$, there is a positive integer N so that $|f_n(x) - f(x)| < \frac{\epsilon}{b-a}$ for all $n > N$ and for all $x \in [a, b]$. Then

$$\left| \int_a^b f_n(x)\, dx - \int_a^b f(x)\, dx \right| = \left| \int_a^b (f_n(x) - f(x))\, dx \right|$$

$$\leq \int_a^b |f_n(x) - f(x)|\, dx$$

$$\leq \int_a^b \frac{\epsilon}{b-a}\, dx$$

$$= \frac{\epsilon}{b-a} \cdot (b-a) = \epsilon \quad \blacksquare$$

You should supply the reason for each of the steps above by citing an appropriate lemma from Section 25.

COROLLARY 30.4   If $f_n(x)$ is integrable on $[a, b]$ for each $n \in \mathbb{N}$, and $\lim_{n \to \infty} f_n(x) = f(x)$, but $\lim_{n \to \infty} \int_a^b f_n(x)\, dx \neq \int_a^b f(x)\, dx$, then the sequence $\{f_n(x)\}$ does not converge uniformly to $f(x)$ on $[a, b]$.

Proof.   This is left as an exercise.

Example 2   Find $\lim_{n \to \infty} \int_a^\pi \frac{\sin nx}{nx}\, dx$ for $a > 0$.

Solution.   To ensure that each f_n is continuous at 0, we first define

$$f_n(0) = \lim_{x \to 0} \frac{\sin nx}{nx} = 1$$

Then
$$f(x) = \lim_{n \to \infty} \frac{\sin nx}{nx} = \begin{cases} 0 & \text{for } x > 0 \\ 1 & \text{for } x = 0 \end{cases}$$

For $\epsilon > 0$, there is some N so that $\frac{1}{Na} < \epsilon$. Since for $n > N$ and for $x \geq a$,

$$\left| \frac{\sin nx}{nx} - 0 \right| \leq \frac{1}{nx} \leq \frac{1}{na} < \epsilon$$

the sequence $\{\frac{\sin nx}{nx}\}$ converges uniformly to 0 on $[a, \pi]$, and so

$$\lim_{n \to \infty} \int_a^\pi \frac{\sin nx}{nx}\, dx = \int_a^\pi 0\, dx = 0$$

by Theorem 30.3. Since f is not continuous at $x = 0$, the convergence cannot be uniform on the interval $[0, \pi]$, so we cannot use Theorem 30.3 to find $\lim_{n \to \infty} \int_0^\pi \frac{\sin nx}{nx}\, dx$. $\square$

After observing the past two theorems in which we proved that if $\{f_n\}$ converged uniformly to f on $[a, b]$, then

$$\lim_{n \to \infty} \lim_{x \to x_o} f_n(x) = \lim_{x \to x_o} \lim_{n \to \infty} f_n(x)$$

and
$$\lim_{n\to\infty} \int_a^b f_n(x)\,dx = \int_a^b \lim_{n\to\infty} f_n(x)\,dx$$

we might be tempted to conjecture also that

$$\lim_{n\to\infty} \frac{d}{dx} f_n(x) = \frac{d}{dx} \lim_{n\to\infty} f_n(x)$$

The next example shows that this conjecture is not always true.

Example 3   Let $f_n(x) = \frac{\sin nx}{n}$ be defined on $\mathbf{R} = (-\infty, \infty)$. The sequence $\{f_n\}$ converges uniformly to $f = 0$ on $\mathbf{R}$, since for all x and for all $n > \frac{1}{\epsilon}$

$$\left| \frac{\sin nx}{n} - 0 \right| \leq \frac{1}{n} < \epsilon$$

However, the sequence $\{f_n'(x)\} = \{\cos nx\}$ does not even converge for all values of x. So $f'(x) \neq \lim f_n'(x)$ for values of x such as $x = 0$, $x = 1$, $x = \frac{\pi}{2}$, or $x = \pi$. $\square$

Theorem 30.6 below asserts that this is a true conclusion however if we change the hypotheses. We first use Theorem 30.3 to obtain a restricted case of this result. If each $f_n'(t)$ is integrable on $[a, b]$, then for $x \in [a, b]$, we have

$$\lim_{n\to\infty} \int_a^x f_n'(t)\,dt = \int_a^x \lim_{n\to\infty} f_n'(t)\,dt$$

by Theorem 30.3. It is also true that

$$\int_a^x f_n'(t)\,dt = f_n(x) - f_n(a)$$

by the Fundamental Theorem of Calculus. If $\lim f_n$ is represented by f, and if $\lim f_n'$ is represented by g, we can put these results together to obtain $f(x) - f(a) = \int_a^x g(t)\,dt$. You should supply the reasons. If we know that g is continuous on $[a, b]$, then we can use Theorem 27.4 to obtain the final result that $f' = g$, which is equivalent to the statement that $\frac{d}{dx} \lim_{n\to\infty} f_n(x) = \lim_{n\to\infty} \frac{d}{dx} f_n(x)$. The above comments furnish an outline of a proof for the following theorem.

THEOREM 30.5 Let the sequence $\{f_n(x)\}$ converge to $f(x)$ on $[a, b]$, and for all n, let $f_n'(x)$ exist and be continuous on $[a, b]$. If the sequence $\{f_n'(x)\}$ converges uniformly to some function $g(x)$ on $[a, b]$, then $f'(x)$ exists on $[a, b]$ and equals $g(x)$.

Proof. This is left as an exercise.

Actually, Theorem 30.6 shows that the same conclusion as in Theorem 30.5 can be obtained from a weaker set of hypotheses, but this requires a more complicated proof.

THEOREM 30.6 Let $\{f_n\}$ be defined on $[a, b]$ and let there be some point $x_o \in [a, b]$ so that the sequence of constants $\{f_n(x_o)\}$ converges. If f_n' exists for all n, and the sequence $\{f_n'\}$ converges uniformly to g on $[a, b]$, then $\{f_n\}$ converges uniformly on $[a, b]$ to a function f so that $f' = g$.

Proof. This can be found on pages 301-303 in *Introduction to Real Analysis* by Robert Bartle and Donald Sherbert.

Example 3 and Theorem 30.6 show that it is possible for $\{f_n\}$ to converge uniformly when $\{f_n'\}$ does not converge uniformly, but that if $\{f_n'\}$ converges uniformly, $\{f_n\}$ must also, except in the case when $\{f_n\}$ does not converge for any value of x.

In Chapter 9, we will apply these important results to a special kind of sequence of functions, that of the sequence of partial sums for a sequence of functions. The most familiar examples of such "series of functions" are the Maclaurin series, the Taylor series, and the Fourier series of functions.

EXERCISES 30

1. Use Corollary 30.2 to make a statement about the uniform convergence of the following sequences.

   (a) $f_n(x) = \frac{x^n}{n}$

   (b) $f_n(x) = \frac{x^n}{1+x^n}$

   (c) $f_n(x) = nxe^{-nx}$

2. If $f(x) = \lim_{n \to \infty} \frac{nx^n}{1+nx^n}$, prove that $f(x) = 0$ for $x \in (-1, 1)$, and $f(x) = 1$ for all other values of x. What does this imply about the uniform convergence of this sequence?

3. Use Theorem 30.3 to prove that $\lim_{n\to\infty} \int_1^2 e^{-nx^2} dx = 0$.

4. Find a value for $\lim_{n\to\infty} \int_{.5}^2 \frac{(\ln x)^n}{n} dx$. Include remarks about uniform convergence as needed.

5. Let $f_n(x) = \frac{nx}{1+nx}$.

   (a) Prove that $\lim \int_0^1 f_n(x)dx = \int_0^1 \lim f_n(x)dx$

   (b) Explain why neither Theorem 30.3 nor Corollary 30.4 applies to the problem in part (a).

6. (a) Find $\lim_{n\to\infty} \int_0^1 \frac{2nx}{1+n^2x^4} dx$.

   (b) Why does Theorem 30.3 not apply to this problem?

7. Let $f_n(x) = \frac{e^{-nx}}{n}$ for $x \geq 0$.

   (a) Find $f = \lim f_n$ and $g = \lim f_n'$.

   (b) Does Theorem 30.6 apply to this sequence?

8. Prove Corollary 30.2.

9. In the proof of Theorem 30.3, prove that $-L(f, P_\epsilon) \leq -L(f_N, P_\epsilon) + \frac{\epsilon}{3}$.

10. Supply the reasons for the steps in the proof of the second part of Theorem 30.3.

11. Prove Corollary 30.4.

12. Organize the hints given in the text to write a proof for Theorem 30.5 using the step-reason format.

13. (a) In what way does the proof of Theorem 30.1 resemble the proof of Theorem 27.3?

    (b) In what way does the proof of the second part of Theorem 30.3 resemble the proof of Theorem 27.3?

8

Series of Constants

SECTION 31   SERIES WITH POSITIVE TERMS

This chapter may be considered as a continuation of Section 9, where we introduced sequences of partial sums. For reference, we repeat the definition and four results that were obtained in that section.

Definition 1   The infinite series $\sum x_n$ *converges to* x if the sequence of partial sums $\{S_n\}$, where $S_n = x_1 + x_2 + \cdots + x_n$, converges to x. Recall that $\sum x_n$ means the same as $\sum_{n=1}^{\infty} x_n$. The series $\sum x_n$ *diverges* if the sequence of partial sums $\{S_n\}$ diverges. For a series of positive terms, the only way that a series can diverge is if $\lim S_n = +\infty$. In this case, we say that the series $\sum x_n$ diverges to $+\infty$.

Result 1   The geometric series $\sum_{n=0}^{\infty} r^n$ converges to $\frac{1}{1-r}$ if $0 < |r| < 1$ and diverges if $|r| \geq 1$.

Result 2   The p-series $\sum_{n=1}^{\infty} \frac{1}{n^p}$ converges if $p > 1$, and diverges if $p \leq 1$.

Result 3   If $\sum_{n=1}^{\infty} x_n$ converges, then $\lim_{n \to \infty} x_n = 0$. Alternatively, if $\lim_{n \to \infty} x_n \neq 0$, then $\sum_{n=1}^{\infty} x_n$ diverges.

Result 4   If $\sum_{n=1}^{\infty} y_n$ converges, and $0 < x_n \leq y_n$ for all n, then $\sum_{n=1}^{\infty} x_n$ converges. In addition, if $\sum_{n=1}^{\infty} y_n$ converges to y, and $\sum_{n=1}^{\infty} x_n$ converges to x, then $x \leq y$. If $\sum_{n=1}^{\infty} y_n$ diverges to $+\infty$, and $0 \leq y_n \leq x_n$ for all n, then $\sum_{n=1}^{\infty} x_n$ diverges to $+\infty$ also.

It is helpful in subsequent proofs and problems to use the following lemma, which you are asked to prove as an exercise.

217

LEMMA 31.1   If x_n is defined for $n \geq 1$, then the following three statements are equivalent.

   (a) The series $\sum_{n=1}^{\infty} x_n$ converges.

   (b) The series $\sum_{n=N}^{\infty} x_n$ converges, for any integer $N > 0$.

   (c) The series $\sum_{n=1}^{\infty} kx_n$ converges, for any constant $k \neq 0$.

Proof.   This is left as an exercise.

You are also asked in the exercises to prove the form of the Comparison Test below, where the hypotheses are adjusted to read that for some $N > 1$, we have $0 < x_n \leq y_n$ for all $n > N$.

THEOREM 31.2 (Comparison Test)

If the series $\sum y_n$ converges and there is some positive integer N so that $0 < x_n \leq y_n$ for all $n > N$, then $\sum x_n$ converges. If the series $\sum y_n$ diverges to $+\infty$, and $0 < y_n \leq x_n$ for all $n > N$, then $\sum x_n$ diverges to $+\infty$ also.

We now prove three well-known convergence tests for positive series which are all derived from Lemma 31.1, and the basic comparison test as expressed in Theorem 31.2.

THEOREM 31.3 (Limit Comparison Test)

Let $\sum x_n$ and $\sum y_n$ be series of positive terms. If $\lim_{n \to \infty} \frac{x_n}{y_n} = L$ where $0 < L < \infty$, then $\sum x_n$ converges if and only if $\sum y_n$ converges.

Proof.   Since there must be a value for N such that

$$\frac{L}{2} < \frac{x_n}{y_n} < \frac{3L}{2}$$

for all $n > N$, it follows that

$$\frac{L}{2} y_n < x_n < \frac{3L}{2} y_n$$

So if $\sum_{n=1}^{\infty} x_n$ converges, then $\sum_{n=N}^{\infty} x_n$ does also by Lemma 31.1, and so does $\sum_{n=N}^{\infty} \frac{L}{2} y_n$ by Theorem 31.2. In turn, $\sum_{n=N}^{\infty} y_n$ and $\sum_{n=1}^{\infty} y_n$ also converge by Lemma 31.1.

On the other hand, if $\sum_{n=1}^{\infty} y_n$ converges, so do $\sum_{n=N}^{\infty} y_n$, $\sum_{n=N}^{\infty} \frac{3L}{2} y_n$, $\sum_{n=N}^{\infty} x_n$ and $\sum_{n=1}^{\infty} x_n$ in that order. You should supply the reasons for each step. ■

Example 1   Because $\sum_{n=1}^{\infty} \frac{1}{n^2}$ is known to converge and

$$\lim_{n \to \infty} \left(\frac{1}{\sqrt{n^4 + n^2 + 2}} \div \frac{1}{n^2} \right) = \lim_{n \to \infty} \frac{n^2}{\sqrt{n^4 + n^2 + 2}} = 1$$

we know that $\sum_{n=1}^{\infty} \frac{1}{\sqrt{n^4+n^2+2}}$ converges by the Limit Comparison Test. Alternatively, we can use Theorem 31.2 and the fact that $\frac{1}{\sqrt{n^4+n^2+2}} < \frac{1}{n^2}$ for all $n \in \mathbf{N}$.   □

If $\lim_{n \to \infty} \frac{x_n}{y_n}$ equals 0 or ∞ in Theorem 31.3, the above proof does not apply, but a partial conclusion can be made, as shown in the two theorems stated below.

THEOREM 31.4 (Limit Comparison Test for Convergence)
Let $\sum x_n$ and $\sum y_n$ be series of positive terms. If $\sum y_n$ converges and $\lim_{n \to \infty} \frac{x_n}{y_n} = L$ where $0 \le L < \infty$, then $\sum x_n$ converges also.

Proof.   The proof in Theorem 31.3 applies if $\lim_{n \to \infty} \frac{x_n}{y_n}$ exists and is greater than 0. If $\lim_{n \to \infty} \frac{x_n}{y_n} = 0$, then there is some value for N so that $\frac{x_n}{y_n} < 1$, or $x_n < y_n$, for $n > N$, and the conclusion follows.   ∎

An intuitive feeling is helpful here. If $\lim \frac{x_n}{y_n} = 0$, this means that x_n is "much smaller" than y_n for large values of n, and hence the convergence of $\sum y_n$ implies the convergence of $\sum x_n$. On the other hand, $\lim \frac{x_n}{y_n} = \infty$ implies that x_n is "much larger" than y_n, so if $\sum y_n$ converges, nothing can be said about the convergence of $\sum x_n$.

THEOREM 31.5 (Limit Comparison Test for Divergence)
Let $\sum x_n$ and $\sum y_n$ be series of positive terms. If $\sum y_n$ diverges and $\lim_{n \to \infty} \frac{x_n}{y_n} = L$, where $0 < L \le \infty$, then $\sum x_n$ diverges also.

Proof.   This proof is left as an exercise.

If one suspects that a given series $\sum x_n$ converges, then the series $\sum \frac{1}{n^2}$ is often a good choice for the comparison series. If one suspects that a given series $\sum x_n$ diverges, then $\sum \frac{1}{n}$ is often a good choice for the comparison series. If both of these fail, as they do in Example 5 below, we need to look for a "larger" convergent series than $\sum \frac{1}{n^2}$ $\left(\sum \frac{1}{n^p} \text{ for } 1 < p < 2 \text{ is a good} \right.$

choice), or a "smaller" divergent series than $\sum \frac{1}{n}$ $\left(\sum \frac{1}{n \ln n}\right.$ is a possible choice). The series $\sum a_n$ is considered to be a "larger" convergent series than $\sum \frac{1}{n^2}$ if $\sum a_n$ converges and $\lim_{n \to \infty}(a_n \div \frac{1}{n^2}) = +\infty$. The series $\sum b_n$ is considered a "smaller" divergent series than $\sum \frac{1}{n}$ if $\sum b_n$ diverges and $\lim_{n \to \infty}(b_n \div \frac{1}{n}) = 0$.

The above comments indicate that the p-series are often a good choice to use as the known series in the Limit Comparison Test. We next look at a test that uses the geometric series as the known comparison series. The proof of the Ratio Test clearly reveals the use of this comparison series, but it is not obvious from the statement of the Ratio Test that the geometric series is involved.

THEOREM 31.6 (Ratio Test)
Let the series of positive terms $\sum x_n$ be given, and assume that $\lim_{n \to \infty} \frac{x_{n+1}}{x_n}$ exists.

If $\lim_{n \to \infty} \frac{x_{n+1}}{x_n} < 1$, then $\sum_{n=1}^{\infty} x_n$ converges.

If $\lim_{n \to \infty} \frac{x_{n+1}}{x_n} > 1$, then $\sum_{n=1}^{\infty} x_n$ diverges.

If $\lim_{n \to \infty} \frac{x_{n+1}}{x_n} = 1$, the test fails.

Proof. If $\lim_{n \to \infty} \frac{x_{n+1}}{x_n} = t < 1$, then for $r = \frac{1+t}{2} < 1$, the geometric series $\sum r^n$ converges. In addition, since $t < r$, we have $\frac{x_{n+1}}{x_n} < r$, or $x_{n+1} < r x_n$, if $n > N$, for some positive integer N. Thus we obtain the following set of inequalities.

$$x_{N+1} < r x_N$$
$$x_{N+2} < r x_{N+1} < r^2 x_N$$
$$\vdots$$
$$x_{N+n} < r^n x_N$$

This implies that $\sum x_n$ converges, using Lemma 31.1 and Theorem 31.2, and so the first part of the theorem is proven. We are now ready to prove the second statement in the theorem. If $\lim_{n \to \infty} \frac{x_{n+1}}{x_n} > 1$, then there is some N so that $x_{n+1} > x_n$ for all $n > N$. But if $x_n \geq x_N > 0$ for $n > N$, then $\lim_{n \to \infty} x_n \neq 0$, and so $\sum x_n$ diverges by the nth term test. · ∎

Remark. The final part of Theorem 31.6 states that the Ratio Test fails if $\lim_{n \to \infty} \frac{x_{n+1}}{x_n} = 1$. When we say that a test fails under this condition, we mean that it is possible to find a convergent series $\sum x_n$ for which $\lim_{n\to\infty} \frac{x_{n+1}}{x_n} = 1$, and a divergent series $\sum y_n$ for which $\lim_{n\to\infty} \frac{y_{n+1}}{y_n} = 1$. For the Ratio Test, $\sum \frac{1}{n^2}$ and $\sum \frac{1}{n}$ serve as examples of such a convergent and divergent series respectively.

Example 2   The series $\sum \frac{2^n}{n!}$ converges by the Ratio Test because

$$\lim_{n\to\infty} \frac{x_{n+1}}{x_n} = \lim_{n\to\infty} \frac{2^{n+1}}{(n+1)!} \cdot \frac{n!}{2^n} = \lim_{n\to\infty} \frac{2}{n+1} = 0 < 1 \quad \square$$

One nice property of the Ratio Test is that it is not necessary to find a series to use for comparison purposes. It is an extremely useful method when we try to find convergence sets for series of functions in Chapter 9. One might reasonably ask if there is another test that uses the p-series as a basis for comparison in the proof, but not in the statement of the test, similar to the way the geometric series is used in the proof of the Ratio Test. Raabe's Test is such a test. While its proof is more involved than the one for the Ratio Test, it is accessible to the reader of this book. Raabe's Test is commonly used on series for which the Ratio Test fails (see Example 3). The test is stated here without proof.

THEOREM 31.7 (Raabe's Test)
Let $x_n > 0$ for all n, and let $\lim_{n\to\infty} n\left(1 - \left(\frac{x_{n+1}}{x_n}\right)\right)$ exist and be equal to t. Then the series $\sum x_n$ converges if $t > 1$ and diverges if $t < 1$.

Proof.   See the reference below.[1]

Example 3   In order to test the series

$$\frac{1}{2} + \frac{1\cdot 3}{2\cdot 4} + \frac{1\cdot 3\cdot 5}{2\cdot 4\cdot 6} + \cdots + \frac{1\cdot 3\cdot 5\cdots(2n-1)}{2\cdot 4\cdot 6\cdots(2n)} + \cdots$$

we first observe that the Ratio Test fails because

$$\lim_{n\to\infty} \frac{x_{n+1}}{x_n} = \lim_{n\to\infty} \frac{2n+1}{2n+2} = 1$$

However,

$$\lim_{n\to\infty} n\left(1 - \frac{x_{n+1}}{x_n}\right) = \lim_{n\to\infty} n\left(1 - \frac{2n+1}{2n+2}\right) = \lim_{n\to\infty} \frac{n}{2n+2} = \frac{1}{2}$$

and so the series diverges by Raabe's Test. $\square$

---

[1] *Advanced Calculus* by Watson Fulks, John Wiley, 1969, pages 392-393.

Neither the Limit Comparison Test nor the Ratio Test can be used to determine the limit value of a convergent series. Assuming that we can evaluate the appropriate improper integral, the Integral Test is perhaps the best general method for approximating the value of a convergent series of positive terms. The proof of this test depends on the idea of comparison, this time using area under a curve.

THEOREM 31.8 (Integral Test)

Let f be a positive, continuous and decreasing function on the interval $[k, \infty)$ and let the sequence $\{x_n\}$ be defined by $x_n = f(n)$. Then the series $\sum x_n$ converges if and only if the improper integral $\int_k^\infty f(x)\,dx$ converges. In addition, the inequalities

$$S_{k-1} + \int_k^\infty f(x)\,dx \;<\; \sum_{n=1}^\infty x_n \;<\; S_k + \int_k^\infty f(x)\,dx$$

give upper and lower bounds for the value of the convergent series.

Proof.   The proof of this test depends on the comparison of the area under the graph of f with the area of the two rectangles shown in Figure 31.1.

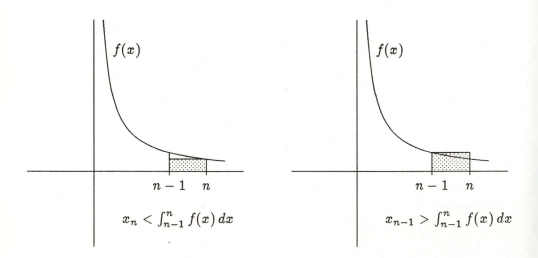

Figure 31.1

From this diagram we see that

$$f(n) \cdot 1 \;<\; \int_{n-1}^{n} f(x)\,dx \;<\; f(n-1) \cdot 1$$

$$x_n \;<\; \int_{n-1}^{n} f(x)\,dx \;<\; x_{n-1} \tag{31.1}$$

Summing over the left inequality in formula (31.1), we have

$$\sum_{n=k+1}^{\infty} x_n \;<\; \sum_{n=k+1}^{\infty} \int_{n-1}^{n} f(x)\,dx \;=\; \int_{k}^{\infty} f(x)\,dx$$

$$\sum_{n=1}^{\infty} x_n \;<\; S_k + \sum_{n=k+1}^{\infty} x_n \;=\; S_k + \int_{k}^{\infty} f(x)\,dx$$

So if $\int_{k}^{\infty} f(x)\,dx$ converges, then the series $\sum x_n$ is bounded above and is therefore convergent, and to a value less than $S_k + \int_{k}^{\infty} f(x)\,dx$.

Summing over the right inequality in formula (31.1), we obtain

$$\sum_{n=k+1}^{\infty} \int_{n-1}^{n} f(x)\,dx \;<\; \sum_{n=k+1}^{\infty} x_{n-1}$$

$$\int_{k}^{\infty} f(x)\,dx \;<\; \sum_{n=k}^{\infty} x_n \tag{31.2}$$

$$S_{k-1} + \int_{k}^{\infty} f(x)\,dx \;<\; S_{k-1} + \sum_{n=k}^{\infty} x_n \;=\; \sum_{n=1}^{\infty} x_n \tag{31.3}$$

If $\int_{k}^{\infty} f(x)\,dx$ diverges to $+\infty$, then so does $\sum_{n=k}^{\infty} x_n$ because of the inequality in formula (31.2).

If $\sum_{n=1}^{\infty} x_n$ converges, then we can use the inequality in formula (31.3) to get

$$S_{k-1} + \int_{k}^{\infty} f(x)\,dx$$

as a lower bound for the value of the series, with the term x_k being a bound on the error of the approximation. This completes the proof of the theorem. ∎

Example 4   We can use the Integral Test to show that $\sum \frac{1}{n^{\frac{3}{2}}}$ converges

because $\int_1^\infty \frac{1}{x^{\frac{3}{2}}} \, dx$ converges to 2, or to show that $\sum \frac{1}{n \ln n}$ diverges because

$$\int_2^\infty \frac{1}{x \ln x} \, dx = \lim_{t \to \infty} \left(\ln (\ln x) - \ln (\ln 2) \right) = \infty$$

You are asked in the exercises to find upper and lower bounds for the
convergent series $\sum \frac{1}{n^{\frac{3}{2}}}$.   □

Example 5   The *p*-series $\sum_{n=1}^\infty \frac{1}{n^p}$ converges if $p > 1$ because

$$\int_1^\infty \frac{1}{x^p} \, dx = \lim_{t \to \infty} \int_1^t x^{-p} \, dx$$

$$= \lim_{t \to \infty} \frac{x^{-p+1}}{-p+1} \bigg|_1^t$$

$$= \lim_{t \to \infty} \frac{1}{1-p} \left(\frac{1}{t^{p-1}} - 1 \right) = \frac{1}{p-1} \quad □$$

Example 6   Apply an appropriate test to decide whether the series
$\sum_{n=2}^\infty \frac{\ln n}{n^2}$ converges or diverges. See also Exercise 28.1a.

Solution.   To apply the Ratio Test, we use the fact that $\lim_{n \to \infty} \frac{n}{n+1} = 1$
and $\lim_{n \to \infty} \frac{\ln(n+1)}{\ln n} = 1$ (by L'Hospital's Rule) to show that

$$\lim_{n \to \infty} \frac{x_{n+1}}{x_n} = \lim_{n \to \infty} \left(\frac{\ln(n+1)}{(n+1)^2} \cdot \frac{n^2}{\ln n} \right)$$

$$= \lim_{n \to \infty} \frac{\ln(n+1)}{\ln n} \cdot \lim_{n \to \infty} \left(\frac{n}{n+1} \right)^2 = 1$$

and therefore the Ratio Test fails for this series.

In order to use the Limit Comparison Test, we could compare $\sum x_n$
with $\sum \frac{1}{n^2}$ if we suspect convergence, and with $\sum \frac{1}{n}$ if we suspect divergence. However

$$\lim_{n \to \infty} \frac{\ln n}{n^2} \cdot \frac{n^2}{1} = \infty \quad \text{and} \quad \lim_{n \to \infty} \frac{\ln n}{n^2} \cdot \frac{n}{1} = 0$$

so the Limit Comparison Test fails (see Theorems 31.4 and 31.5) for both
these comparison series. In this case, we need to use a convergent series

that is "greater" than $\sum \frac{1}{n^2}$, or a divergent series that is "less" than $\sum \frac{1}{n}$. The series $\sum \frac{1}{n^{\frac{3}{2}}}$ meets the first condition, while $\sum \frac{1}{n \ln n}$ meets the second. Since

$$\lim_{n \to \infty} \frac{\ln n}{n^2} \cdot \frac{n^{\frac{3}{2}}}{1} = \lim_{n \to \infty} \frac{\ln n}{n^{\frac{1}{2}}} = 0$$

by L'Hospital's Rule, $\sum \frac{\ln n}{n^2}$ converges by Theorem 31.4. $\square$

Example 7 Find an approximation for the limit value of the convergent series in Example 6.

Solution. Since the terms are all positive, we can use the Integral Test to find bounds for this value. Using integration by parts,

$$\int_k^\infty \frac{\ln x}{x^2} = \lim_{t \to \infty} \left(\frac{-\ln x}{x} \Big|_k^t + \int_k^t \frac{1}{x^2} dx \right)$$

$$= \lim_{t \to \infty} \left(\frac{-\ln t}{t} + \frac{\ln k}{k} - \frac{1}{t} + \frac{1}{k} \right) = \frac{1 + \ln k}{k}$$

The Integral Test requires that $\frac{\ln x}{x^2}$ is a decreasing function on $[k, \infty)$. We observe that $\frac{\ln x}{x^2}$ has its relative maximum at $(\sqrt{e}, \frac{1}{2e})$, so $k \geq 3$ must hold in order to use the Integral Test. By the Integral Test, we obtain the approximations shown below.

$$S_2 + \int_3^\infty \frac{\ln x}{x^2} dx < \sum_{n=2}^\infty \frac{\ln n}{n^2} < S_3 + \int_3^\infty \frac{\ln x}{x^2} dx$$

$$\frac{\ln 2}{4} + \frac{1 + \ln 3}{3} < \sum_{n=2}^\infty \frac{\ln n}{n^2} < \frac{\ln 2}{4} + \frac{\ln 3}{9} + \frac{1 + \ln 3}{3}$$

$$.872824 < \sum_{n=2}^\infty \frac{\ln n}{n^2} < .994893 \quad \square$$

It is left as an exercise to show that this approximation can be generalized to the one shown in formula (31.4) below, which is valid for all $k \geq 3$.

$$S_{k-1} + \frac{1 + \ln k}{k} < \sum_{n=2}^\infty \frac{\ln n}{n^2} < S_k + \frac{1 + \ln k}{k} \qquad (31.4)$$

We now summarize the results derived in this section. There are three main methods for testing a positive series $\sum x_n$ for convergence.

Method 1. If an antiderivative can be found for the function f, where $f(n) = x_n$ and f is decreasing on $[k, \infty)$, then the Integral Test may be the best method to use. If the series converges, then $S_k + \int_k^\infty f(x)\,dx$ serves as an upper bound for the value of the series, while $S_{k-1} + \int_k^\infty f(x)\,dx$ serves as a lower bound.

Method 2. Usually one can find $\lim_{n\to\infty} \frac{x_{n+1}}{x_n}$ quite easily. However, it often happens that this limit is 1, in which case the Ratio Test fails, and so Raabe's Test may be tried. But if $\lim_{n\to\infty} \frac{x_{n+1}}{x_n} = L \neq 1$, the convergence or divergence of $\sum_{n=1}^\infty x_n$ is settled, although no information is given about the value for this convergent series. If $\lim_{n\to\infty} \frac{x_{n+1}}{x_n}$ does not exist, then there are other forms of the Ratio Test, but these are more difficult to apply.

Method 3. We can routinely test $\sum x_n$ with $\sum \frac{1}{n}$ and $\sum \frac{1}{n^2}$ using the Limit Comparison Test. If these both fail, then we must try a "second level" of known series such as $\sum \frac{1}{n \ln n}$ and $\sum \frac{1}{n^{\frac{3}{2}}}$. If these also fail, we must go to a third level, or to another test.

For most series we encounter, at least one of these three main tests will work. If not, there is a collection of other tests that can be applied. These include the Root Test, Gauss's Test, Kummer's Test, or adjusted forms of the Ratio Test. All of these are well-known results and can be found in the literature if needed.

EXERCISES 31

1. What p-series would you use to test the series $\sum \frac{4+n^2}{\sqrt{n^5+2n^3-7}}$ by the Limit Comparison Test? Show how to apply the Limit Comparison Test.

2. (a) Use $\int_1^\infty \frac{1}{x^{\frac{3}{2}}}\,dx$ to find an upper bound and a lower bound for the series
$$1 + \frac{1}{2\sqrt{2}} + \frac{1}{3\sqrt{3}} + \cdots .$$

   (b) Use $\int_2^\infty \frac{1}{x^{\frac{3}{2}}}\,dx$ to find a more accurate upper bound for $\sum_{n=1}^\infty \frac{1}{n^{\frac{3}{2}}}$.

   (c) Prove that the sequence $\{A_n\} = \{S_n + \int_n^\infty \frac{1}{x^{\frac{3}{2}}}\,dx\}$ is a decreasing sequence of upper bounds for the series $\sum \frac{1}{n^{\frac{3}{2}}}$.

(d) Find an increasing sequence $\{B_n\}$ of lower bounds for the series $\sum \frac{1}{n^{\frac{3}{2}}}$
that uses the integral $\int_n^\infty \frac{1}{x^{\frac{3}{2}}}\, dx$.

3. Test $\sum \frac{\ln n}{n^{\frac{3}{2}}}$ for convergence.

4. Test $\sum_{n=2}^\infty \frac{1}{n \ln n}$ for convergence.

5. Find a divergent series $\sum b_n$, so that $\lim_{n \to \infty}(b_n \div \frac{1}{n}) = 0$.

6. Which of the following conditions guarantees the convergence of $\sum x_n$, when $x_n > 0$ for each n. Explain your reasoning

   (a) $\frac{1}{n^{\frac{3}{2}}} < x_n < \frac{1}{n}$ holds for all n.

   (b) $\frac{x_{n+1}}{x_n} < 1$ holds for all n.

   (c) $\lim_{n \to \infty} n^2 x_n = 0$.

7. Show that $\lim_{n \to \infty} \frac{1}{n^2} \div r^n = \infty$ for $0 < r < 1$. This means that no geometric series can be used to test $\sum \frac{1}{n^2}$ for convergence.

8. Give some examples of "subseries" of the harmonic series $\sum \frac{1}{n}$ that converge, and some examples that diverge. Can you find a property that characterizes the convergent subseries?

9. Prove that $f(x) = \frac{\ln x}{x^2}$ has a relative maximum at $\left(\sqrt{e}, \frac{1}{2e}\right)$, as stated in Example 5.

10. Use Definition 1 in order to prove Lemma 31.1.

11. Prove Theorem 31.2.

12. Supply reasons for the proof of Theorem 31.3.

13. Prove Theorem 31.5. Look at the form of the proof for Theorem 31.4.

14. Use Raabe's Test to show that the following series converges.

$$\sum \frac{1 \cdot 3 \cdot 5 \cdots (2n-1)}{2 \cdot 4 \cdot 6 \cdots (2n)} \cdot \frac{1}{2n+1}$$

In Chapter 9, we will see that this series converges to arcsin 1.

15. Verify the inequality in formula 31.4 after Example 7.

SECTION 32 SERIES WITH MIXED SIGNS

We turn now to a consideration of series with mixed signs, meaning that they have both an infinite number of positive and negative terms. Since the sequence $\{S_n = x_1 + \cdots + x_n\}$ is no longer increasing, none of the tests in Section 31 is available for our use. A helpful insight can be obtained by forming S_n for one of the best known such series,

$$\sum_{n=1}^{\infty} \frac{(-1)^{n+1}}{n} = 1 - \frac{1}{2} + \frac{1}{3} - \frac{1}{4} + - \cdots$$

Since by definition, this series does whatever the sequence of partial sums does, we begin with the observation that the first ten terms of $\{S_n\}$ are 1, .5, .833, .583, .783, .616, .758, .633, .744, and .644.

The subsequence of the odd subscripted terms

$$\{S_{2n-1}\} = \{1, .833, .783, .758, .744, \cdots\}$$

is decreasing, while the subsequence of the even subscripted terms

$$\{S_n\} = \{.5, .583, .616, .633, .644, \cdots\}$$

is increasing. Since both subsequences are bounded, they must converge, and we only need to show that they converge to the same value (see Corollary 7.6). This general technique is used in the following test, which can be applied whenever the series alternates in sign.

THEOREM 32.1 (Alternating Series Test)

If the series $\sum_{n=1}^{\infty} (-1)^{n+1} x_n$ has the property that $0 < x_{n+1} < x_n$ and $\lim_{n \to \infty} x_n = 0$, then $\sum_{n=1}^{\infty} (-1)^{n+1} x_n$ converges, and to a value S so that $|S - S_n| < x_{n+1}$.

Proof. The key to this proof is to show that

$$S_2 < S_4 < \cdots < S_{2n} < \cdots < S_{2n+1} < \cdots < S_3 < S_1 \quad (32.1)$$

From this, it will follow that $\{S_{2n}\}$ and $\{S_{2n+1}\}$ are monotone and bounded sequences, and must therefore converge. Since $S_{2n+1} = S_{2n} + x_{2n+1}$ and $\lim_{n \to \infty} x_{2n+1} = 0$, it follows that $\lim_{n \to \infty} S_{2n+1} = \lim_{n \to \infty} S_{2n}$. Because the subsequences, both of the odd subscripted terms and the even subscripted terms, converge to the same value, say S, the sequence $\{S_n\}$ must also converge to S by Corollary 7.6. And since $S_n < S < S_{n+1}$ for even values of n and $S_{n+1} < S < S_n$ for odd values of n, it follows that

$$|S - S_n| < |S_{n+1} - S_n| < x_{n+1}$$

The key inequality in equation (32.1) above can be established from the following three observations.

(a) $S_{2n} = S_{2n-2} + (x_{2n-1} - x_{2n}) > S_{2n-2}$

(b) $S_{2n+1} = S_{2n-1} - x_{2n} + x_{2n+1} = S_{2n-1} - (x_{2n} - x_{2n+1}) < S_{2n-1}$

(c) $S_{2n+1} = S_{2n} + x_{2n+1} > S_{2n}$ ■

Example 1 The series $\sum_{n=1}^{\infty}(-1)^n \frac{\ln n}{n^2}$ converges by the Alternating Series Test, because the sequence $\{\frac{\ln n}{n^2}\}$ is decreasing, and its limit is 0, as was shown in Example 31.5. □

While the Alternating Series Test is very easy to use, it often does not apply. In addition to the Alternating Series Test, there is a second general test for a series with mixed signs. It is called the **Absolute Convergence Test** and is proved by adapting the Cauchy sequence concept from Section 8 to a series setting (see Exercise 8.3). This result is the first important application of the concept of a Cauchy sequence. The second application will be the Weierstrass M-Test in Section 34.

THEOREM 32.2 (Cauchy Convergence Criterion)
The series $\sum_{n=1}^{\infty} x_n$ converges if and only if for every $\epsilon > 0$, there is some N so that for $n > m > N$, it follows that $\left|\sum_{i=m+1}^{n} x_i\right| < \epsilon$.

Proof. The proof follows by showing that the following five statements are equivalent. You are asked to verify this as an exercise.

(1) The series $\sum_{n=1}^{\infty} x_n$ converges.

(2) $\{S_n\} = \{x_1 + x_2 + \cdots + x_n\}$ is a convergent sequence.

(3) $\{S_n\}$ is a Cauchy sequence.

(4) For each $\epsilon > 0$, there is some N, so that $|S_n - S_m| < \epsilon$, for all $n, m \in \mathbf{N}$ for which $n > m > N$.

(5) For $\epsilon > 0$, there is some N so that $|x_{m+1} + x_{m+2} + \cdots + x_n| < \epsilon$ if $n > m > N$. ■

THEOREM 32.3 (Absolute Convergence Test)

If the series $\sum_{n=1}^{\infty} |x_n|$ converges, then the series $\sum_{n=1}^{\infty} x_n$ converges also.

Proof.   Let $\epsilon > 0$ be given. By the above criterion, $\sum x_n$ converges if there is some N, so that $\left| \sum_{i=m+1}^{n} x_i \right| < \epsilon$ for all $n > m > N$.

But since $\sum |x_n|$ converges, there is some N so that for $n > m > N$, we have $\left| \sum_{i=m+1}^{n} |x_i| \right| < \epsilon$. Then

$$\left| \sum_{i=m+1}^{n} x_i \right| \leq \sum_{i=m+1}^{n} |x_i| = \left| \sum_{i=m+1}^{n} |x_i| \right|$$

and the proof is complete. ■

The Absolute Convergence Test provides another means to test a series $\sum x_n$ with mixed signs. We can apply the Comparison Test, the Ratio Test, the Integral Test, or any other test for a series of positive terms to the series $\sum |x_n|$. If this series converges, then the original series $\sum x_n$ must also converge. If the series of absolute values diverges, then no conclusion can be made about the convergence of $\sum x_n$.

Example 2   Apply the Alternating Series Test and the Absolute Convergence Test to the following series with mixed signs.

(a) $1 + \frac{1}{3} - \frac{1}{2} + \frac{1}{5} + \frac{1}{7} - \frac{1}{4} + + - \cdots$

(b) $\sum \frac{\sin n}{n^2}$

(c) $1 + \frac{1}{4} - \frac{1}{9} + \frac{1}{16} + \frac{1}{25} - \frac{1}{36} + + - \cdots$

(d) $1 + \frac{1}{2} - \frac{1}{3} - \frac{1}{4} + \frac{1}{5} + \frac{1}{6} - - + + \cdots$

(e) $1 - \frac{1}{9} + \frac{1}{4} - \frac{1}{36} + \frac{1}{16} - \frac{1}{81} + \frac{1}{25} - \frac{1}{144} + - \cdots$

(f) $\sum \frac{\sin n}{n}$

Solution.

(a) By arranging the terms of this series in groups of three as

$$\left(1 + \frac{1}{3} - \frac{1}{2} \right) + \left(\frac{1}{5} + \frac{1}{7} - \frac{1}{4} \right) + \cdots$$

we can show that the general term is

$$\frac{1}{4n - 3} + \frac{1}{4n - 1} - \frac{1}{2n} = \frac{8n - 3}{(4n - 3)(4n - 1)(2n)}$$

which converges by comparison with $\sum \frac{1}{n^2}$. If $\{S_n\}$ represents the sequence of partial sums, the above argument shows that the subsequence $\{S_{3n}\}$ converges. An additional argument is needed to prove that $\{S_n\}$ converges.

(b) Since $\left|\frac{\sin n}{n^2}\right| \le \frac{1}{n^2}$, the series $\sum_{n=1}^{\infty} \left|\frac{\sin n}{n^2}\right|$ converges as a series of positive terms by comparison with $\sum \frac{1}{n^2}$. By the Absolute Convergence Test, $\sum \frac{\sin n}{n^2}$ converges also.

(c) The Absolute Convergence Test once again gives convergence.

(d) By grouping, we can make this into the alternating series

$$\left(1 + \frac{1}{2}\right) - \left(\frac{1}{3} + \frac{1}{4}\right) + \left(\frac{1}{5} + \frac{1}{6}\right) + \cdots + (-1)^{n+1} \left(\frac{4n-1}{4n^2 - 2n}\right) + \cdots$$

and since the sequence $\left\{\frac{4n-1}{4n^2-2n}\right\}$ decreases to 0, the given series converges by the Alternating Series Test. This verifies that $\{S_{2n}\}$ converges, and once again, an additional argument is needed to prove that $\{S_n\}$ converges. You can do this as an exercise by showing that the subsequence $\{S_{2n-1}\}$ converges, and then proving that $\{S_{2n}\}$ and $\{S_{2n-1}\}$ converge to the same value.

(e) This series is a rearrangement of the terms in (c) and we need the material on rearrangements below before we can decide on convergence or divergence.

(f) None of the tests or methods described above apply to this series. □

Definition 1   The series $\sum x_n$ is said to *converge absolutely* if the series $\sum |x_n|$ converges. The series $\sum x_n$ is said to *converge conditionally* if $\sum x_n$ converges and $\sum |x_n|$ diverges. Note that if $\sum x_n$ diverges, then $\sum |x_n|$ must also diverge.

Example 3   The series $\sum \frac{1}{n^2}$ converges, so $\sum \frac{(-1)^{n+1}}{n^2}$ converges absolutely, while $\sum \frac{(-1)^{n+1}}{n}$ converges conditionally since $\sum \frac{1}{n}$ diverges. □

For the remainder of this section, while we consider the difference between absolute and conditional convergence, we continue to assume that

$\sum x_n$ has an infinite number both of positive and negative terms (see Exercise 7). The proofs of the next two theorems are relatively simple, if you are not confused by the symbolism that we now introduce. While there are different approaches to these proofs, each involves a fair amount of notation. We let $\sum p_n$ represent the series of the positive terms in $\sum x_n$, taken in the order in which they occur. In a similar way, let $\sum q_n$ represent the series of absolute values of the negative terms, taken in the order in which they occur.

For example, in the series of Example 2a, where

$$\sum x_n = 1 + \frac{1}{3} - \frac{1}{2} + \frac{1}{5} + \frac{1}{7} - \frac{1}{4} + + - \cdots$$

we have
$$\sum p_n = 1 + \frac{1}{3} + \frac{1}{5} + \frac{1}{7} + \cdots$$

and
$$\sum q_n = \frac{1}{2} + \frac{1}{4} + \frac{1}{6} + \cdots$$

We are therefore working with the four infinite series $\sum x_n$, $\sum |x_n|$, $\sum p_n$, and $\sum q_n$. We use $\{S_n\}$, $\{T_n\}$, $\{P_n\}$, and $\{Q_n\}$ respectively to represent the sequences of partial sums for these series. We let S, T, P, and Q represent the limit values of these sequences (i.e. if $\sum x_n$ converges, then $S = \sum_{n=1}^{\infty} x_n$, and if $\sum |x_n|$ diverges, then $T = +\infty$). Finally, for a given value of n, let i_n denote the number of positive terms and j_n denote the number of negative terms from the first n terms of the series $\sum_{n=1}^{\infty} x_n$. It therefore follows that

$$\begin{aligned} S_n &= P_{i_n} - Q_{j_n} \\ T_n &= P_{i_n} + Q_{j_n} \\ n &= i_n + j_n \end{aligned}$$

Example 4   As an example of this notation, for the series

$$\sum x_n = 1 + \frac{1}{3} - \frac{1}{2} + \frac{1}{5} + \frac{1}{7} - \frac{1}{4} + \frac{1}{9} + \frac{1}{11} - \frac{1}{6} + + - \cdots$$

we have

$$S_7 = 1 + \frac{1}{3} - \frac{1}{2} + \frac{1}{5} + \frac{1}{7} - \frac{1}{4} + \frac{1}{9}$$

$$= \left(1 + \frac{1}{3} + \frac{1}{5} + \frac{1}{7} + \frac{1}{9}\right) - \left(\frac{1}{2} + \frac{1}{4}\right)$$

$$= P_5 - Q_2$$

Also $S_{17} = P_{12} - Q_5$, as you should show. $\square$

THEOREM 32.4 If $\sum x_n$ converges absolutely to S, then $\sum p_n$ and $\sum q_n$ both converge. In this case, $S = P - Q$ and $T = P + Q$. If $\sum x_n$ converges conditionally to S, then $\sum p_n$ and $\sum q_n$ both diverge to $+\infty$.

Proof. If the series $\sum x_n$ converges absolutely to S, then the series $\sum |x_n|$ also converges (to a value we denote by T). This means that the sequence $\{T_n\}$ is increasing and convergent with limit value T. Since $T_n = P_{i_n} + Q_{j_n}$ and $T_n \le T$, we have $P_{i_n} \le T_n \le T$ and $Q_{j_n} \le T_n \le T$.

Since $\{P_n\}$ and $\{Q_n\}$ are monotone and bounded sequences, they both converge, which is equivalent to saying that the series $\sum p_n$ and $\sum q_n$ converge. Furthermore,

$$S = \lim_{n \to \infty} S_n = \lim_{n \to \infty} (P_{i_n} - Q_{j_n})$$

$$= \lim_{n \to \infty} P_{i_n} - \lim_{n \to \infty} Q_{i_n} = P - Q$$

and similarly, $T = P + Q$.

If $\sum x_n$ converges conditionally to S, then $\{S_n\}$ converges to S and $\{T_n\}$ diverges to ∞. Since $P_{i_n} = \frac{1}{2}(S_n + T_n)$ and $Q_{j_n} = \frac{1}{2}(T_n - S_n)$, it follows that $\{P_n\}$ and $\{Q_n\}$ both diverge to ∞. ∎

Definition 2 A series $\sum y_n$ is called a *rearrangement* of the series $\sum x_n$ if it contains all the same terms as $\sum x_n$ but in a different order. This means that for each term (i.e. y_m) in $\sum y_n$, there is some term (i.e. x_n) in $\sum x_n$ so that $y_m = x_n$, and conversely. This can be stated more formally by saying there is a 1-1 function p from **N** onto **N**, so that $y_n = x_{p(n)}$ for each $n \in \mathbf{N}$.

Example 5   The series in Example 2a is a rearrangement of the alternating harmonic series $\sum_{n=1}^{\infty} \frac{(-1)^{n+1}}{n}$, and the series in Example 2e is a rearrangement of the series $\sum_{n=1}^{\infty} \frac{(-1)^{n+1}}{n^2}$.   □

Absolute and conditional convergence are strikingly contrasted by observing their effects on rearrangements of series, as described in the following theorem. The proof uses the notation that was developed for Theorem 32.4.

THEOREM 32.5 (Theorem on Rearrangements)

If $\sum x_n$ converges absolutely to S, then all rearrangements of $\sum x_n$ converge to S also. If $\sum x_n$ converges conditionally to S, then a given rearrangement could converge or diverge. In fact, for any real number c, there is a rearrangement of $\sum x_n$ that converges to c.

Proof.   We first consider the series $\sum x_n$ of positive terms and then let $S_n = x_1 + \cdots + x_n$. Let $\sum y_n$ be an arbitrary rearrangement of $\sum x_n$, and let $Y_n = y_1 + \cdots + y_n$. Since every y_i value is one of the terms in $\sum x_n$, for every n there is some value m so that $Y_n \leq S_m \leq S$, which implies that $\sum y_n$ converges, and to a value Y, so that $Y \leq S$.

Now that we know $\sum y_n$ converges, we can reverse the roles of $\sum x_n$ and $\sum y_n$ above to obtain $S \leq Y$. Therefore the rearrangement series $\sum y_n$ converges to S.

For a general series $\sum x_n$ with mixed signs that converges absolutely, we have $\sum x_n = \sum p_{i_n} + \sum q_{j_n}$, where $\sum x_n$ converges to S, $\sum p_{i_n}$ converges to P, and $\sum q_{j_n}$ converges to Q by Theorem 32.4. A rearrangement series $\sum y_n$ will have the series of positive terms $\sum \overline{p}_{i_n}$ and the series of the absolute values of the negative terms $\sum \overline{q}_{j_n}$. Since $\sum \overline{p}_{i_n}$ and $\sum \overline{q}_{j_n}$ are rearrangements of the series of positive terms $\sum p_{i_n}$ and $\sum q_{j_n}$, they must converge to P and Q respectively. Therefore $\sum y_n$ converges to $P - Q = S$.

On the other hand, if $\sum x_n$ converges conditionally, then both $\sum p_n$ and $\sum q_n$ must diverge to $+\infty$. Thus to obtain a rearrangement that will converge to an arbitrary positive real number c, we first choose enough terms from $\sum p_n$ until the sum exceeds c. Next, we subtract off terms in order from $\sum q_n$ until the new sum is less than c. Continuing this process, we will obtain a rearrangement that converges to c. Since $\sum p_n$ and $\sum q_n$ both diverge to ∞, we will never run short of terms to carry out the procedure described above. A similar process would apply if $c \leq 0$, or even if $c = \pm\infty$. Riemann was the first to observe this property. ∎

Example 6   In order to obtain a rearrangement of the alternating harmonic series that converges to -1, we proceed as follows. We first observe that $-\frac{1}{2} - \frac{1}{4} - \frac{1}{6} - \frac{1}{8} = -1.041\overline{6}$. Then we add in $p_1 = 1$ to obtain the new sum of $-.041\overline{6}$. Going back to the negative terms, we continue until the new sum is again less than -1, and then add in $p_2 = \frac{1}{3}$, and continue this process. The rearranged series that converges to -1 will begin as follows.

$$-\frac{1}{2} - \frac{1}{4} - \frac{1}{6} - \frac{1}{8} + 1 - \frac{1}{10} - \frac{1}{12} - \frac{1}{14} - \cdots - \frac{1}{62} + \frac{1}{3} - \frac{1}{64} - \frac{1}{66} - \cdots \quad \square$$

EXERCISES 32

1. Test $\sum \frac{\sin n}{n^{\frac{3}{2}}}$ for convergence.

2. Test $1 + \frac{1}{2} - \frac{1}{3} + \frac{1}{4} + \frac{1}{5} - \frac{1}{6} + + - \cdots$ by the grouping approach illustrated in Example 2a.

3. Test $\sum \frac{(-1)^n \ln n}{n}$ by the Alternating Series Test. How can you establish the decreasing property?   Consider Corollary 20.5.

4. Show that $1 - \frac{1}{2} - \frac{1}{4} + \frac{1}{3} - \frac{1}{6} - \frac{1}{8} + - - \cdots = \frac{1}{2}\left(1 - \frac{1}{2} + \frac{1}{3} - \frac{1}{4} + - \cdots\right)$.
   This gives a rearrangement of the alternate harmonic series that converges to a different value than $\ln 2$.

5. Can the series $1 - \frac{1}{2\sqrt{2}} + \frac{1}{3\sqrt{3}} - \frac{1}{4\sqrt{4}} + - \cdots$ be rearranged to converge to 2? Explain.

6. Write out the first 40 terms of the rearrangement of the alternating harmonic series $\sum_{n=1}^{\infty} \frac{(-1)^{n+1}}{n}$ that converges to $\sqrt{2}$.

7. Explain why the distinction between absolute and conditional convergence is meaningless unless the series has an infinite number both of positive and negative terms.

8. Explain how you could form a rearrangement of the alternating harmonic series that diverges to $+\infty$.

9. Supply reasons for each of the steps in the proof of Theorem 32.2.

10. Why does the Absolute Convergence Test not apply to Example 2f, as it did to Example 2b?

11. What can be said about a series of mixed signs $\sum x_n$, if $\sum p_n$ converges to P, and $\sum q_n$ diverges to ∞? Show that your answer implies that the two statements in Theorem 32.4 are really if and only if statements.

12. Verify the remark in Example 2d that the sequence $\left\{ \frac{4n-1}{4n^2-2n} \right\}$ decreases to 0.

13. Use Theorem 32.5 to decide whether the series in Example 2e converges or diverges.

14. How many terms of $\sum_{n=2}^{\infty} \frac{(-1)^n}{\ln n}$ are needed to obtain an approximation for the value of this series with an error less than .01?

15. Complete the work to prove that the sequence in Example 2d converges.

SECTION 33 VALUES OF CONVERGENT SERIES

All the convergence tests in the previous two sections are existence theorems in the following sense. The series $\sum_{n=1}^{\infty} x_n$ converges if and only if there *exists* a real number x so that for each $\epsilon > 0$, there is some N so that

$$\left| \sum_{i=1}^{n} x_i - x \right| < \epsilon \qquad \text{for } n > N$$

While the tests do not usually reveal what the value is for x, the following comments summarize the methods we have for finding approximate or exact values for x. For series with positive terms, the terms of the sequence of partial sums $\{S_n\}$ provide lower bounds for x, while the Integral Test gives $S_n + \int_n^{\infty} f(x)\,dx$ as upper bounds. For series with alternating signs, the terms of the sequence $\{S_n\}$ furnish both upper and lower bounds in an alternating pattern. These are perhaps the only general results for approximating x, except for Maclaurin series representation of functions, which are discussed in the next chapter.

When it comes to finding exact values for x, the geometric series is probably the best general method that we have. However, there are special methods that can sometimes be used to find an "exact" value for x. In this setting, $\frac{\pi^2}{6}$ will be considered an exact value, even though we can only approximate it by a decimal representation.

Method 1 The geometric series $\sum_{n=0}^{\infty} r^n$ converges to the value $\frac{1}{1-r}$ if $0 < |r| < 1$. For example,

$$\sum_{n=1}^{\infty} \frac{1}{2^n} = \frac{1}{2} \cdot \frac{1}{1 - \frac{1}{2}} = 1$$

Method 2 The exact value of a series such as

$$\sum_{n=1}^{\infty} \frac{1}{(n+a)(n+b)(n+c)}$$

where the general term is a rational function whose denominator has only non-repeated linear factors, can be found by the method of partial fractions. You encountered this in calculus as a method of integration. For the series $\sum \frac{1}{n(n+3)}$, we use partial fractions to write

$$\frac{1}{n(n+3)} = \frac{1}{3}\left(\frac{1}{n}\right) - \frac{1}{3}\left(\frac{1}{n+3}\right)$$

so that

$$\sum_{n=1}^{\infty} \frac{1}{n(n+3)} = \frac{1}{3}\left(1 - \frac{1}{4}\right) + \frac{1}{3}\left(\frac{1}{2} - \frac{1}{5}\right) + \frac{1}{3}\left(\frac{1}{3} - \frac{1}{6}\right) + \cdots$$

$$= \frac{1}{3}\left(1 + \frac{1}{2} + \frac{1}{3}\right) = \frac{11}{18}$$

Method 3 In the section on Maclaurin series representation of functions coming up in the next chapter, we can substitute a specific value into the Maclaurin series to find a value for the limit of a sequence. For example, by letting $x = 1$ in the Maclaurin series for e^x, we see that

$$\sum_{n=1}^{\infty} \frac{1}{n!} \quad \text{converges to} \quad e - 1$$

and

$$\sum_{n=1}^{\infty} \frac{(-1)^{n+1}}{2n - 1} \quad \text{converges to} \quad \frac{\pi}{4}$$

by letting $x = 1$ in the Maclaurin series for $\arctan x$.

Method 4 We will see in Section 36 how to use the Fourier series expansion of x^2 in order to show that $\sum_{n=1}^{\infty} \frac{1}{n^2} = \frac{\pi^2}{6}$.

Method 5 Sometimes it is possible to find a formula containing the values of two convergent series, so that when the value of one is found, the other is also known. As an example, let

$$S = \sum_{n=1}^{\infty} \frac{1}{n^2} = 1 + \frac{1}{4} + \frac{1}{9} + \frac{1}{16} + \cdots$$

$$T = \sum_{n=1}^{\infty} \frac{(-1)^{n+1}}{n^2} = 1 - \frac{1}{4} + \frac{1}{9} - \frac{1}{16} + - \cdots$$

Then

$$S - T = 2\left(\frac{1}{4} + \frac{1}{16} + \frac{1}{36} + \cdots\right)$$

$$= \frac{1}{2}\left(1 + \frac{1}{4} + \frac{1}{9} + \cdots\right) = \frac{1}{2}S$$

and so $T = \frac{S}{2}$.

Thus any approximation of T by the Alternating Series Test gives an approximation for S, or any approximation of S by the Integral Test gives an approximation for T. And if we are lucky enough to find an exact value for S, as we can using Fourier Series, then we also have an exact value for T.

We come now to the second part of this section, which is to discuss values for divergent series. You may feel that this is a meaningless question, because in calculus we tend to think of divergent series as useless and disposable. Once we determined that a series diverged, we had no more to do with it. However, divergent series have been studied as a serious topic during the past 150 years, and a few comments are now included about them. Morris Kline, in his book *Mathematical Thought from Ancient to Modern Times*, devotes Chapter 47 to this interesting idea, while the number theorist from Cambridge University, G. H. Hardy, wrote an entire book on this topic.[1] He considered this his best book, although you may be more familiar with his little book called *A Mathematician's Apology*, in which Hardy discussed his reasons for being a mathematician.

It was found that a number of series that diverged in the formal sense gave very good approximate solutions to some physical problems if only a relatively few terms were used. The French mathematician Henri Poincare called such divergent series *asymptotic*, and spent a fair amount of his research on this idea. It should not be too surprising that some divergent series were used in the eighteenth century for applications, since there was no firm concept of a distinction between convergent and divergent series in that century. It was the occurrence of some paradoxical results about convergence that could not be resolved by intuition that led to the formal definition of convergence in the early nineteenth century.

There is a concept of summability that uses an averaging process under which some divergent series can be shown to be summable to a finite value. For example, the divergent series $1 - 1 + 1 - 1 + 1 - 1 + - \cdots$ can be shown to be summable to $\frac{1}{2}$. This idea was presented in Section 7.

---

[1] *Divergent Series* by G. H. Hardy, Oxford: The Clarendon Press, 1949.

EXERCISES 33

1. Find the value for these series.

    (a) $3 + \frac{6}{5} + \frac{12}{25} + \frac{24}{125} + \cdots$

    (b) $1 - \frac{x}{2} + \frac{x^2}{4} - \frac{x^3}{8} + - \cdots$

    (c) $1 + \frac{1}{3} + \frac{1}{5} + \frac{1}{7} + \cdots$

2. Use the method of partial fractions to prove that

$$\sum_{n=1}^{\infty} \frac{1}{n(n+1)(n+2)} = \frac{1}{4}$$

3. Prove that for $p \in \mathbb{N}$,

$$\sum_{k=1}^{\infty} \frac{1}{k(k+1)\cdots(k+p)} \quad \text{converges to} \quad \frac{1}{p \cdot p!}$$

4. (a) If S is the limit value for the convergent p-series $\sum_{n=1}^{\infty} \frac{1}{n^3}$, and T is the limit value for the convergent alternating series $\sum_{n=1}^{\infty} \frac{(-1)^{n+1}}{n^3}$, prove that $T = \frac{3S}{4}$.

    (b) Similarly, if V is the value for the convergent series $\sum_{n=1}^{\infty} \frac{1}{(2n-1)^3}$, prove that $V = \frac{7S}{8}$. Thus any approximation for the value S gives an approximation for the values of T and V.

5. Find an approximate value for S in Exercise 4 with accuracy to three decimal places.

6. Find an approximate value for T in Exercise 4 with accuracy to three decimal places by two different methods. One method is to use the results in Exercises 4 and 5. A second method is to use the Alternating Series Test.

7. Prove that the series

$$1 + \frac{1}{3} - \frac{1}{2} + \frac{1}{5} + \frac{1}{7} - \frac{1}{4} + \frac{1}{9} + \frac{1}{11} - \frac{1}{6} + + - \cdots$$

converges to $\frac{3}{2} \ln 2$, given that the alternating harmonic series

$$1 - \frac{1}{2} + \frac{1}{3} - \frac{1}{4} + \frac{1}{5} - \frac{1}{6} + - \cdots$$

converges to $\ln 2$. See also Example 32.2a.

9

Series of Functions

SECTION 34 THEOREMS ON UNIFORM CONVERGENCE

In Chapter 7, we proved three theorems about the interchange of limits
for a sequence of functions. Since the pointwise and uniform convergence
for an infinite series of functions is defined in terms of the pointwise and
uniform convergence of the sequence of partial sums of the series, we will
have three corresponding theorems for infinite series of functions. Because
of the widespread usage of Taylor series and Fourier series in analysis,
these theorems have important applications. Their proofs follow directly
from those in Chapter 7. You may want to review Chapter 7 to help you
understand the translation of those results about sequences of functions
to the setting of series of functions in this chapter. We begin with the
definition of pointwise and uniform convergence for a series of functions.

Definition 1   The series of functions $\sum_{n=1}^{\infty} u_n(x)$ *converges to* $u(x)$
on the set S if the corresponding sequence of partial sums of functions
$\{f_n(x)\} = \{\sum_{i=1}^{n} u_i(x)\}$ converges to $u(x)$ on S in the sense of Defini-
tion 29.1. Observe that $u(x) = \lim_{n\to\infty} f_n(x) = \lim_{n\to\infty} \sum_{i=1}^{n} u_i(x) = \sum_{i=1}^{\infty} u_i(x)$. Similarly, $\sum_{n=1}^{\infty} u_n(x)$ *converges uniformly on* S if the sequence $\{f_n(x) = \sum_{i=1}^{n} u_i(x)\}$ converges uniformly on S in the sense of
Definition 29.2.

Example 1   The geometric series $\sum_{n=0}^{\infty} x^n$ converges to $\frac{1}{1-x}$ on $(-1, +1)$
because the sequence

$$\left\{ \sum_{i=0}^{n} x^i \right\} = \left\{ \frac{1 - x^{n+1}}{1 - x} \right\}$$

converges to $\frac{1}{1-x}$ on $(-1, +1)$. The convergence is not uniform however
on $(-1, +1)$.   □

THEOREM 34.1   If the series $\sum_{n=1}^{\infty} u_n(x)$ converges uniformly to $u(x)$ on S, and each $u_n(x)$ is continuous on S, then $u(x)$ is continuous on S.

Proof.   Since each $u_n(x)$ is continuous on S, so is $f_n(x) = \sum_{i=1}^{n} u_i(x)$ by Exercise 15.7. Since $\sum_{n=1}^{\infty} u_n(x)$ converges uniformly to $u(x)$ on S, this means that $\{f_n(x)\}$ converges uniformly to $u(x)$ on S also. By Theorem 30.1, $u(x)$ is continuous on S.   ∎

The following example illustrates how we can interchange the processes of integration and summation. For now, we assume that the series in Example 2 converges uniformly, and that the result in Theorem 30.2 is available for our use.

Example 2   Evaluate $\int_0^{\pi} \sum_{n=1}^{\infty} \frac{\sin nx}{n^2}\, dx$.

Solution.   It will follow that

$$\int_0^{\pi} \sum_{n=1}^{\infty} \frac{\sin nx}{n^2}\, dx \;=\; \sum_{n=1}^{\infty} \int_0^{\pi} \frac{\sin nx}{n^2}\, dx$$

by Theorem 30.2, since we can show in Example 4 that the series $\sum_{n=1}^{\infty} \frac{\sin nx}{n^2}$ converges uniformly on the set $[0, \pi]$. Then since

$$\int_0^{\pi} \frac{\sin nx}{n^2}\, dx \;=\; \frac{-\cos nx}{n^3}\Big|_0^{\pi} \;=\; \frac{1 - (-1)^n}{n^3}$$

we have from Exercise 33.4b,

$$\int_0^{\pi} \sum_{n=1}^{\infty} \frac{\sin nx}{n^2}\, dx \;=\; \sum_{n=1}^{\infty} \frac{1 - (-1)^n}{n^3}$$

$$=\; \frac{2}{1^3} + \frac{0}{2^3} + \frac{2}{3^3} + \frac{0}{4^3} + \cdots$$

$$=\; 2 \sum_{n=1}^{\infty} \frac{1}{(2n-1)^3}$$

$$=\; \frac{7}{4} \sum_{n=1}^{\infty} \frac{1}{n^3} \;\approx\; 2.10358 \quad \square$$

We now prove a general theorem that gives sufficient conditions for the interchange of the processes of integration and summation.

THEOREM 34.2   If the series $\sum_{n=1}^{\infty} u_n(x)$ converges uniformly to the function $u(x)$ on a set S, and if each $u_n(x)$ is integrable on $[a, b] \subseteq S$, then $u(x)$ is also integrable on $[a, b]$, and

$$\int_a^b u(x)\, dx \;=\; \int_a^b \sum_{n=1}^{\infty} u_n(x)\, dx \;=\; \sum_{n=1}^{\infty} \int_a^b u_n(x)\, dx$$

Proof.   We apply Theorem 30.2 to the sequence of functions $\{f_n(x)\}$ where $f_n(x) = u_1(x) + u_2(x) + \cdots + u_n(x)$. Each f_n is integrable on $[a, b]$ by Lemma 25.3, so

$$\lim_{n\to\infty} f_n(x) \;=\; \lim_{n\to\infty} \sum_{i=1}^{n} u_i(x) \;=\; \sum_{i=1}^{\infty} u_i(x) \;=\; u(x)$$

is also integrable on $[a, b]$.   Also

$$\int_a^b u(x)\, dx \;=\; \int_a^b \lim_{n\to\infty} f_n(x)\, dx$$

$$=\; \lim_{n\to\infty} \int_a^b f_n(x)\, dx \qquad \text{by Theorem 30.2}$$

$$=\; \lim_{n\to\infty} \int_a^b \sum_{i=1}^{n} u_i(x)\, dx$$

$$=\; \lim_{n\to\infty} \sum_{i=1}^{n} \int_a^b u_i(x)\, dx \qquad \text{by Lemma 25.3}$$

$$=\; \sum_{i=1}^{\infty} \int_a^b u_i(x)\, dx \;=\; \sum_{n=1}^{\infty} \int_a^b u_n(x)\, dx \qquad \blacksquare$$

The next theorem describes sufficient conditions for the interchange of the processes of differentiation and summation.

THEOREM 34.3   If $\sum_{n=1}^{\infty} u_n(x)$ converges for some value x_o in the interval J, and if $\sum_{n=1}^{\infty} u_n'$ converges uniformly on J, then $\sum_{n=1}^{\infty} u_n(x)$ converges uniformly to a function $u(x)$ on J, and also $u'(x)$ exists and equals $\sum_{n=1}^{\infty} u_n'(x)$.

Proof.   This is left as an exercise.

Example 3   Let the function $f(x)$ be defined by the series

$$\sum_{n=1}^{\infty} nx^n = x + 2x^2 + 3x^3 + \cdots$$

Show that $f(x) = \frac{x}{(1-x)^2}$.

Solution.   We assume for now that x belongs to a set on which the given series converges uniformly (such as $[-r, r]$, for $0 < r < 1$), so that we can apply Theorems 34.1, 34.2, and 34.3. For $x \neq 0$,

$$\frac{f(x)}{x} = \sum_{n=1}^{\infty} nx^{n-1} = 1 + 2x + 3x^2 + \cdots$$

$$\int \frac{f(x)}{x} \, dx = \sum_{n=1}^{\infty} \int nx^{n-1} \, dx = \sum_{n=1}^{\infty} x^n$$

$$= x + x^2 + x^3 + \cdots = \frac{x}{1 - x}$$

Then by differentiating both sides of the above equation, we have

$$\frac{f(x)}{x} = \frac{d}{dx}\left(\frac{x}{1-x}\right) = \frac{1}{(1-x)^2}$$

and so for all x for which $-r \leq x \leq r$,

$$f(x) = \frac{x}{(1-x)^2}$$

Observe that this formula is still correct if $x = 0$.   □

   To consider this problem from the reverse point of view, we could apply Taylor's Theorem to $f(x) = \frac{x}{(1-x)^2}$, and show that the Taylor polynomials are $\sum_{n=1}^{k} nx^n$. It would be difficult to find a pattern for the derivatives $f^{(n)}(0)$ to complete this approach however.

   Since each of the three "interchange of limit" theorems for a series of functions $\sum u_n(x)$ contains the hypothesis of uniform convergence, it is

essential that we know how to determine whether a specific series of functions converges uniformly. In Chapter 7, the only means to determine if a
sequence of functions $\{f_n(x)\}$ converged uniformly was by the definition.
We are fortunate to have a more efficient means to determine if the series
of functions $\sum_{n=1}^{\infty} u_n(x)$ converges uniformly. It is called the Weierstrass
M-Test, and is proved using the Cauchy convergence criterion that was
presented in Section 32.

THEOREM 34.4 (Weierstrass M-Test)

If the series of functions $\sum_{n=1}^{\infty} u_n(x)$ has the property that $|u_n(x)| \leq M_n$
for each n, and for all $x \in S$, where $\sum_{n=1}^{\infty} M_n$ is a convergent series of
positive constants, then $\sum_{n=1}^{\infty} u_n(x)$ converges uniformly on S.

Proof.   We first extend the Cauchy convergence criterion in Theorem
32.2 to assert that the series $\sum_{n=1}^{\infty} u_n(x)$ converges uniformly on S if and
only if for each $\epsilon > 0$, there is some N_ϵ so that if $n > m > N_\epsilon$, then
$|\sum_{i=m+1}^{n} u_i(x)| < \epsilon$ for all $x \in S$.

Now let $\epsilon > 0$ be given. Since $\sum M_n$ converges, there is some N_ϵ so
that $n > m > N_\epsilon$ implies that $|\sum_{i=m+1}^{n} M_n| < \epsilon$. But by the hypotheses
above,

$$\left| \sum_{i=m+1}^{n} u_i(x) \right| \leq \sum_{i=m+1}^{n} |u_i(x)| \leq \sum_{i=m+1}^{n} M_n = \left| \sum_{i=m+1}^{n} M_n \right| < \epsilon$$

and therefore $\sum_{n=1}^{\infty} u_n(x)$ converges uniformly on S.   ∎

Example 4   The series $\sum_{n=1}^{\infty} \frac{\sin nx}{n^2}$ converges uniformly for all x because
$\sum_{n=1}^{\infty} M_n = \sum_{n=1}^{\infty} \frac{1}{n^2}$ converges, and $\left| \frac{\sin nx}{n^2} \right| \leq \frac{1}{n^2}$ for all x. We could
not make the same assertion for the series $\sum_{n=1}^{\infty} \frac{\sin nx}{n}$.   □

Example 5   The series $\sum_{n=1}^{\infty} \frac{x^n}{n}$ converges uniformly for sets of the form
$[-r, r]$, where $r < 1$, because $\sum_{n=1}^{\infty} r^n$ is a convergent geometric series and
$\left| \frac{x^n}{n} \right| \leq |x|^n \leq r^n$ on such sets.   □

EXERCISES 34

1. Apply the Weierstrass M-Test to show that the following series converge uniformly on the given sets.

   (a) $\sum_{n=1}^{\infty} \frac{(\sin x)^n}{n}$ converges uniformly on $[-r, r]$, for $r < \frac{\pi}{2}$

   (b) $\sum_{n=1}^{\infty} (x \ln x)^n$ converges uniformly on $(0, 1]$.

   (c) $\sum_{n=1}^{\infty} ne^{-nx}$ converges uniformly on $[t, \infty)$, for $t > 0$.

2. Use Exercise 1c and properties of geometric series to show that

$$\int_1^2 \sum_{n=1}^{\infty} ne^{-nx}\, dx = \frac{e}{e^2 - 1}$$

   See Example 3.

3. Let $f(x) = \sum_{n=1}^{\infty} \frac{(\sin x)^n}{n}$. Show that $f(x) = \ln \frac{1}{1 - \sin x}$.

4. (a) Show that $\sum_{n=1}^{\infty} \frac{\cos nx}{n^2}$ converges uniformly for all x, and then show that

$$\int_0^{\frac{\pi}{2}} \sum_{n=1}^{\infty} \frac{\cos nx}{n^2}\, dx = \sum_{n=1}^{\infty} \frac{(-1)^{n+1}}{(2n - 1)^3}$$

   (b) Use the Alternating Series Test to find an approximation for this value with an error less than .001.

5. Let $f(x) = \sum_{n=1}^{\infty} \frac{1}{n} e^{\frac{x}{n}}$. Find the values of x for which

$$f'(x) = \sum_{n=1}^{\infty} \frac{d}{dx} \left(\frac{1}{n} e^{\frac{x}{n}} \right)$$

6. Show that $f(x) = \sum_{n=1}^{\infty} n^{-x^2}$ converges for $|x| > 1$, and that this series converges uniformly on the interval $[k, K]$, where $1 < k < K$.

7. Use Theorem 30.6 in order to prove Theorem 34.3.

SECTION 35 POWER SERIES

We use the term *power series* to refer to a series of the form $\sum_{n=1}^{\infty} a_n x^n$ or $\sum_{n=1}^{\infty} a_n (x - c)^n$. The form $\sum a_n x^n$ is called a *Maclaurin series*, while $\sum a_n (x - c)^n$ is called a *Taylor series* in powers of $(x - c)$. The Maclaurin series is a special case of the Taylor series, obtained when $c = 0$, and is the most commonly used form. The Taylor series, in turn, is but a translation of the Maclaurin series with the interval of convergence being centered at c instead of the origin. All the theorems in this section will be proved for series of the form $\sum a_n x^n$, but they can be easily restated for the more general case $\sum a_n (x - c)^n$.

The results of this section are extremely important for applications because they establish the fact that if a function can be represented by a power series expansion of the form $\sum_{n=1}^{\infty} a_n x^n$ using Taylor's Theorem, then the series and its derivatives and integrals will converge uniformly on what we will call the interval of convergence. Therefore, the three main theorems of Section 34 can be applied to such series and functions. After the results in this section are proved, they will apply to all power series expansions, and we will not have to check each series separately for uniform convergence as we did in Section 34. Remembering this goal should keep you from getting lost in the many details needed to establish the above results.

Recall that in our study of sequences of functions in Chapter 7, we proceeded as follows. We first found the set S where the sequence converged pointwise. Then we found those subsets of S where the convergence is uniform. On such sets, we could apply the interchange of limit theorems about continuity, integrability, and differentiability. In a similar way, we now study series of functions $\sum a_n x^n$. The first order of business is to find where a given series converges pointwise. A favorite technique to use is the Ratio Test.

The Ratio Test (Theorem 31.6) asserts that the series $\sum_{n=1}^{\infty} a_n x^n$ converges for all x for which

$$\lim_{n \to \infty} \left| \frac{a_{n+1} x^{n+1}}{a_n x^n} \right| = |x| \cdot \lim_{n \to \infty} \left| \frac{a_{n+1}}{a_n} \right| < 1$$

and diverges if $|x| \cdot \lim_{n \to \infty} \left| \frac{a_{n+1}}{a_n} \right| > 1$. This simple observation provides the idea for the proof of the following theorem.

THEOREM 35.1 (The Interval of Convergence)

Given the series $\sum_{n=1}^{\infty} a_n x^n$, assume that $t = \lim_{n \to \infty} \left| \frac{a_n}{a_{n+1}} \right|$ exists.

If $t = 0$, then $\sum a_n x^n$ converges only for $x = 0$.

If $t = +\infty$, then $\sum a_n x^n$ converges for all x.

If $0 < t < +\infty$, then $\sum a_n x^n$ converges if $|x| < t$, and diverges if $|x| > t$. The question of convergence for $x = \pm t$ must be settled by another test.

Proof.   This is left as an exercise.

Comment.   In every case, the set of pointwise convergence is an interval centered about the origin. If $t = 0$, the interval is $[0, 0]$. If $t = +\infty$, the interval is $(-\infty, +\infty)$. And if $0 < t < \infty$, the interval will be one of the following: $(-t, t)$, $(-t, t]$, $[-t, t)$, or $[-t, t]$. This convergence set is called the *interval of convergence* for $\sum a_n x^n$, and t is called the *radius of convergence*.

Example 1   Find the interval of convergence for the series

$$\sum a_n x^n = \sum_{n=0}^{\infty} \frac{x^n}{n!}$$

Solution.   The series converges for all x because

$$t = \lim_{n \to \infty} \left| \frac{a_n}{a_{n+1}} \right| = \lim_{n \to \infty} \left| \frac{(n+1)!}{n!} \right| = \lim_{n \to \infty} (n+1) = +\infty$$

Thus, the interval of convergence is $(-\infty, \infty)$. This series is known as the Maclaurin series for e^x.   □

Example 2   Find the interval of convergence for $\sum_{n=0}^{\infty} (-1)^n n! \, x^n$.

Solution.   This series converges only for $x = 0$ because

$$t = \lim_{n \to \infty} \left| \frac{a_n}{a_{n+1}} \right| = \lim_{n \to \infty} \left| \frac{(-1)^n n!}{(-1)^{n+1}(n+1)!} \right| = \lim_{n \to \infty} \frac{1}{n+1} = 0$$

A series that diverges for all $x \neq 0$ may appear to be a rather useless series, but in the 18th century, Euler devised more powerful summability processes to assign "values" to series that diverge according to the definition of convergence we have given.   □

Example 3 Find the interval of convergence for $\sum_{n=1}^{\infty} \frac{x^n}{n2^n}$.

Solution. The series converges if $|x| < 2$ and diverges if $|x| > 2$ since

$$t = \lim_{n \to \infty} \left| \frac{a_n}{a_{n+1}} \right| = \lim_{n \to \infty} \left| \frac{(n+1)2^{n+1}}{n \cdot 2^n} \right| = 2$$

If $x = 2$, the series becomes $\sum \frac{2^n}{n2^n} = \sum \frac{1}{n}$, which diverges to ∞, and if $x = -2$, the series becomes $\sum \frac{(-2)^n}{n2^n} = \sum \frac{(-1)^n}{n}$, which converges by the Alternating Series Test. Thus the interval of convergence is $[-2, 2)$. □

Example 4 Find the convergence set for the power series $\sum_{n=1}^{\infty} \frac{x^{2n+1}}{n+1}$.

Solution. By the Ratio Test, the series will converge if

$$\lim_{n \to \infty} \left| \frac{x^{2n+3}}{n+2} \cdot \frac{n+1}{x^{2n+1}} \right| = x^2 \cdot \lim_{n \to \infty} \frac{n+1}{n+2} = x^2 < 1$$

which occurs when $x \in (-1, 1)$. After testing the endpoints, we finish with $[-1, 1)$ as the interval of convergence. □

Example 5 Find the interval of convergence for the series $\sum_{n=1}^{\infty} \frac{n(x-2)^n}{n+2}$.

Solution. By the Ratio Test, this series will converge when

$$\lim_{n \to \infty} \left| \frac{a_n}{a_{n+1}} \right| = \lim_{n \to \infty} \left| \frac{(n+1)(x-2)^{n+1}}{n+3} \cdot \frac{n+2}{n(x-2)^n} \right|$$

$$= |x-2| \lim_{n \to \infty} \frac{(n+1)(n+2)}{n(n+3)} = |x-2| \cdot 1 < 1$$

which occurs if and only if $1 < x < 3$.

The series diverges at both endpoints by the nth term test, so $(1, 3)$ is the interval of convergence. You might compare this to the series $\sum_{n=1}^{\infty} \frac{nx^n}{n+2}$, which also has a radius of convergence of 1. The interval of convergence $(-1, 1)$ for $\sum \frac{nx^n}{n+2}$ is centered at 0, while the interval of convergence $(1, 3)$ for $\sum \frac{n(x-2)^n}{n+2}$ is centered at 2. □

We next prove a theorem that asserts that a power series converges absolutely on its interval of convergence. You may be tempted to believe that Theorem 35.1 already proves this, because the symmetry of the interval of convergence implies that $\sum a_n x^n$ converges if and only if $\sum a_n(-x)^n$ converges (except possibly at $x = t$). However absolute convergence requires that $\sum |a_n x^n|$ converges, and this is not obvious from Theorem 35.1, if the a_n terms take on negative values. The following proof shows how to handle this situation.

THEOREM 35.2 (Absolute convergence of power series)
If $\sum a_n x^n$ converges for $x = x_o$, where $x_o \neq 0$, then $\sum a_n x^n$ converges absolutely for all x for which $|x| < |x_o|$. If $\sum a_n x^n$ diverges for $x = x_1$, then $\sum a_n x^n$ diverges for all x for which $|x| > |x_1|$.

Proof. If $\sum a_n x_o^n$ converges, then the sequence $\{a_n x_o^n\}$ converges to 0 and is therefore bounded, say $|a_n x_o^n| \leq B$ for all n. Choose x so that $|x| < |x_o|$ and define $r = \left| \frac{x}{x_o} \right|$. Observe that $\sum B r^n$ is a convergent geometric series, since $r = \left| \frac{x}{x_o} \right| < 1$. Then $\sum a_n x^n$ converges absolutely by the comparison test because

$$\left| a_n x^n \right| = \left| a_n x_o^n \frac{x^n}{x_o^n} \right| = \left| a_n x_o^n \right| \cdot \left| \frac{x}{x_o} \right|^n \leq B r^n$$

If $\sum a_n x_1^n$ diverges, then $\sum a_n x^n$ must diverge for all $|x| > |x_1|$. For if $\sum a_n x_2^n$ converged for some x_2 for which $|x_2| > |x_1|$, then $\sum a_n x_1^n$ would have to converge also. Why? ■

THEOREM 35.3 (Uniform Convergence of Power Series)
For $\sum a_n x^n$, let $t = \lim_{n \to \infty} \left| \frac{a_n}{a_{n+1}} \right|$ as before. Then $\sum a_n x^n$ converges uniformly on sets of the form $[-r, r]$, for any r so that $0 < r < t$.

Proof. The series $\sum a_n r^n$ converges absolutely by Theorem 35.2. Therefore $\sum |a_n| r^n$ is a convergent series of positive constants. Since $|a_n x^n| \leq |a_n| r^n$ for all $x \in [-r, r]$, it follows that $\sum a_n x^n$ converges uniformly on $[-r, r]$ by the Weierstrass M-Test. ■

Comment.   There is one case that is not covered by this theorem and that is when the radius of convergence t is finite and $\sum a_n t^n$ converges. In this case, the uniform convergence will be on sets of the form $[-r, t]$, instead of only $[-r, r]$, for $r < t$. This will be very important because we are often interested in the value of a series at an endpoint of the interval of convergence. For example, we can show that intervals of the form $[-r, 1]$, where $0 < r < 1$, are intervals of uniform convergence for the following two series,

$$\arctan x \;=\; x - \frac{1}{3}x^3 + \frac{1}{5}x^5 - \frac{1}{7}x^7 + - \cdots$$

$$\ln(x+1) \;=\; x - \frac{1}{2}x^2 + \frac{1}{3}x^3 - \frac{1}{4}x^4 + - \cdots$$

both of which have $(-1, 1]$ as their interval of convergence. This result was proved by Abel using Dirichlet's Test. It is more difficult to prove than the other theorems in this section—you may consult the reference[1] if interested.

We can summarize the results about pointwise, absolute, and uniform convergence for the series $\sum_{n=1}^{\infty} a_n x^n$ by the following statements.

(1) If the interval of convergence is $[0, 0]$, then absolute convergence and uniform convergence are meaningless.

(2) If the interval of convergence is $(-\infty, \infty)$, then the convergence is absolute on $(-\infty, +\infty)$, and uniform on $[-r, r]$, for any $r \in \mathbf{R}$.

(3) If the interval of convergence is $(-t, t)$ for $t \neq 0$, then the convergence is absolute on $(-t, t)$ and uniform on $[-r, r]$, for all r so that $0 < r < t$.

(4) If the interval of convergence is $[-t, t]$ for $t \neq 0$, then the convergence is absolute and uniform on $[-t, t]$.

(5) If the interval of convergence is $(-t, t]$ for $t \neq 0$, then the convergence is absolute on $(-t, t)$, and uniform on sets of the form $[-r, t]$, for all r so that $0 < r < t$.

(6) If the interval of convergence is $[-t, t)$ for $t \neq 0$, then the convergence is absolute on $(-t, t)$, and uniform on $[-t, r)$, for $0 < r < t$.

---

[1] *Advanced Calculus*, Second Edition, by Watson Fulks, New York: John Wiley and Sons, 1969, page 538.

We next show that the series obtained from $\sum a_n x^n$ by differentiating term-by-term (i.e., $\sum n a_n x^{n-1}$) or integrating term-by-term (i.e., $\sum \frac{a_n}{n+1} x^{n+1}$) will have the same radius of convergence as the original series. In the case where t is finite, it is possible that the interval of convergence might be different at an endpoint. The same type of results about uniform convergence will therefore apply to the series $\sum n a_n x^{n-1}$ and $\sum \frac{a_n}{n+1} x^{n+1}$.

THEOREM 35.4   Let the series $\sum a_n x^n$ have radius of convergence t. Then the series of derivatives $\sum n a_n x^{n-1}$ and the series of integrals $\sum \frac{a_n}{n+1} x^{n+1}$ will also have t as their radius of convergence.

Proof.   Since

$$\sum_{n=0}^{\infty} \frac{d}{dx}(a_n x^n) = \sum_{n=1}^{\infty} n a_n x^{n-1}$$

is again a power series, the radius of convergence for this series of derivatives $\sum \frac{d}{dx}(a_n x^n)$ is

$$\lim_{n \to \infty} \left| \frac{n a_n}{(n+1)a_{n+1}} \right| = \lim_{n \to \infty} \left| \frac{a_n}{a_{n+1}} \right|$$

which is also the radius of convergence for $\sum a_n x^n$. The proof for the case of the series of integrals $\sum_{n=0}^{\infty} \int a_n x^n$ is left as an exercise.   ∎

Example 6   In Example 3, we showed that the interval of convergence for $\sum \frac{x^n}{n 2^n}$ was $[-2, 2)$. The interval of convergence for

$$\sum_{n=1}^{\infty} \frac{d}{dx}\left(\frac{x^n}{n 2^n} \right) = \sum_{n=1}^{\infty} \frac{x^{n-1}}{2^n}$$

is $(-2, 2)$, and the interval of convergence for

$$\sum_{n=1}^{\infty} \int \frac{x^n}{n 2^n} = \sum_{n=1}^{\infty} \frac{x^{n+1}}{n(n+1) 2^n}$$

is $[-2, 2]$. While the radius of convergence is the same for all three series, the situation at the endpoints differs with each one.   □

The following summary statement ties together the results in Sections 21, 29, 30, and 34, and gives the theoretical basis for the applications of power series. The general problem is this.

> Given a function f with derivatives of all orders at $c = 0$, find the coefficients a_n and the set S, so that the series $\sum_{n=1}^{\infty} a_n x^n$ converges uniformly to f on S. Then numerical values associated with the function f may be calculated using the processes of term-by-term differentiation and integration.

The steps in the solution of this problem are as follows.

(1) The coefficients are expressed by the formula

$$a_n = \frac{f^{(n)}(0)}{n!}$$

as is seen in the statement of Taylor's Theorem, or by the intuitive approach outlined at the end of Section 21. A discussion of how the a_n terms are found in specific cases is given in Example 2.11.

(2) The set S on which $\sum a_n x^n$ converges pointwise can be found by the Ratio Test (see Theorem 35.1). This set will always be an interval centered at the origin.

(3) In step (2), we showed that the series converges on S. In order to show that it converges to f on S, we must show that the remainder term, as expressed in Taylor's Theorem, converges to 0 for all $x \in S$. This is usually the most difficult step.

(4) In order to apply the processes of term-by-term differentiation and integration, it is necessary to establish uniform convergence on subsets of S. The theory in this section provides a complete answer to this question. Since the $u_n(x)$ are polynomials, it is clear that each $u_n(x)$ is continuous, differentiable, and integrable everywhere.

(5) It is helpful to obtain a listing of the Maclaurin series for the familiar functions studied in calculus, as a reference for future work. These can be obtained one by one, working through the above steps for each function.

A Table of Maclaurin Series

$$(1) \quad \sin x \ = \ x - \frac{x^3}{3!} + \frac{x^5}{5!} - \frac{x^7}{7!} + \cdots + (-1)^n \frac{x^{2n+1}}{(2n+1)!} + \cdots$$

with interval of convergence $(-\infty, \infty)$

$$(2) \quad \cos x \ = \ 1 - \frac{x^2}{2!} + \frac{x^4}{4!} - \frac{x^6}{6!} + \cdots + (-1)^n \frac{x^{2n}}{(2n)!} + \cdots$$

with interval of convergence $(-\infty, \infty)$

$$(3) \quad \tan x \ = \ x + \frac{x^3}{3} + \frac{2x^5}{15} + \frac{17x^7}{315} + \cdots$$

$$\cdots + \frac{(-1)^{n-1} 2^{2n} (2^{2n} - 1) B_{2n} x^{2n-1}}{(2n)!} + \cdots$$

where B_n is the nth Bernoulli number, and

with interval of convergence $(-\frac{\pi}{2}, \frac{\pi}{2})$

$$(4) \quad e^x \ = \ 1 + x + \frac{x^2}{2!} + \frac{x^3}{3!} + \cdots + \frac{x^n}{n!} + \cdots$$

with interval of convergence $(-\infty, \infty)$

$$(5) \quad \ln(1+x) \ = \ x - \frac{x^2}{2} + \frac{x^3}{3} - \frac{x^4}{4} + \cdots + \frac{(-1)^{n+1} x^n}{n} + \cdots$$

with interval of convergence $(-1, 1]$

$$(6) \quad (1+x)^r \ = \ 1 + rx + \frac{r(r-1)}{2!} x^2 + \frac{r(r-1)(r-2)}{3!} x^3 + \cdots$$

$$\cdots + \frac{r(r-1)\cdots(r-n+1)}{n!} x^n + \cdots$$

for $r \in \mathbf{R}$, and

with interval of convergence $(-1, +1)$

(7) $\quad \dfrac{1}{1+x} \;=\; 1 - x + x^2 - x^3 + x^4 - \cdots + (-1)^n x^n + \cdots$

with interval of convergence $(-1, +1)$

(8) $\quad \arctan x \;=\; x - \dfrac{x^3}{3} + \dfrac{x^5}{5} - \cdots + \dfrac{(-1)^n x^{2n+1}}{2n+1} + \cdots$

with interval of convergence $(-1, 1]$

(9) $\quad \arcsin x \;=\; x + \dfrac{1}{2}\dfrac{x^3}{3} + \dfrac{1 \cdot 3}{2 \cdot 4}\dfrac{x^5}{5} + \cdots$

$$\cdots + \dfrac{1 \cdot 3 \cdots (2n-3)}{2 \cdot 4 \cdots (2n-2)}\dfrac{x^{2n-1}}{2n-1} + \cdots$$

with interval of convergence $[-1, 1]$

(10) $\quad \sinh x \;=\; x + \dfrac{x^3}{3!} + \dfrac{x^5}{5!} + \cdots + \dfrac{x^{2n-1}}{(2n-1)!} + \cdots$

with interval of convergence $(-\infty, \infty)$

(11) $\quad \cosh x \;=\; 1 + \dfrac{x^2}{2!} + \dfrac{x^4}{4!} + \cdots + \dfrac{x^{2n}}{(2n)!} + \cdots$

with interval of convergence $(-\infty, \infty)$

The following comments can be made about the formulas in the above table of Maclaurin series.

1. All these formulas except for (3), (8), and (9) can be obtained by the familiar process of calculating several derivatives and looking for a pattern for $f^{(n)}(0)$. Formula (3) can be obtained from formulas (1) and (2) using long division, while formulas (8) and (9) can be obtained by use of formula (6) and term-by-term integration.

2.   A reason for using the symbols $\sinh x$ and $\cosh x$ for the hyperbolic functions $\sinh x = \frac{e^x - e^{-x}}{2}$ and $\cosh x = \frac{e^x + e^{-x}}{2}$ can be seen by comparing formulas (10) and (11) with formulas (1) and (2).

3. The binomial expansion (6) has a place of central importance for several reasons.

   (a) Newton claimed that its discovery in 1665–66 gave him the insight he needed to "invent" the calculus.

   (b) If r is a natural number, then the series contains only a finite number of non-zero terms. In this case, the coefficients are the well-known binomial coefficients of n things taken m at a time, which you may recognize as the values that appear in Pascal's triangle.

   (c) Setting $r = \frac{1}{2}$ or $\frac{1}{3}$ allows us to find square or cube roots.

   (d) Setting $r = -1$ gives the geometric series in formula (7).

   (e) Setting $r = -\frac{1}{2}$, replacing x by $-x^2$, and integrating term-by-term gives the series for $\arcsin x$ in formula (9). A similar procedure gives the series for $\arctan x$ in formula (8).

Example 7   Find an approximation for the number e.

Solution.   In Section 6, we approximated the value of e by using the convergent sequence $\left\{ \left(1 + \frac{1}{n}\right)^n \right\}$. We can obtain an approximation more easily and accurately with power series. By setting $x = 1$ in the Maclaurin series for e^x, we find $e \approx \sum_{n=1}^{k} \frac{1}{n!}$. This increasing sequence of partial sums provides only lower bounds for e; for example, $e \approx 2.716$ when $k = 5$ and $e \approx 2.71822787$ when $k = 8$.

   In order to find upper bounds, we can set $x = -1$ in the Maclaurin series for e^x, thus obtaining the alternating series $\sum_{n=0}^{\infty} \frac{(-1)^n}{n!}$, which gives both lower and upper bounds for $\frac{1}{e}$. So by taking reciprocals, we can find lower and upper bounds for e. Since

$$\frac{1}{e} = 1 - 1 + \frac{1}{2} - \frac{1}{6} + \frac{1}{24} - \frac{1}{120} + - \cdots$$

we have $.3678571 < \frac{1}{e}$ or $e < 2.7184466$ for $n = 7$.   □

The above argument shows how to obtain a very good approximation for e using a relatively small numbers of terms from the series $\sum \frac{1}{n!}$. An intuitive way to see why this is true is to note that $x = 1$ is much closer to the center $(x = 0)$ of the interval of convergence than to the endpoint of $+\infty$. As a general rule, the closer that x is chosen to the center of the interval (or the furthest that x is chosen from an endpoint), the more rapid will be the rate of convergence. Whenever we must choose x at an endpoint of the interval of convergence, the rate of convergence may be agonizingly slow. Of course, with computers this is far less of a problem for us than it was for mathematicians prior to the twentieth century.

Example 8   Find an approximation for $\ln 2$.

Solution.   Perhaps the most obvious approach is to set $x = 1$ in the series for $\ln(1 + x)$, obtaining the result that

$$\ln 2 = 1 - \frac{1}{2} + \frac{1}{3} - \frac{1}{4} + \frac{1}{5} - \frac{1}{6} + - \cdots$$

This is certainly true, but because $x = 1$ is an endpoint of the interval of convergence, the rate of convergence is slow; 200 terms are needed for 2 decimal place accuracy, 2000 terms for 3 decimal place accuracy, and so on.

Sometimes, an alternate approach is available. If we set $x = -\frac{1}{2}$ in the series for $\ln(1 + x)$, we obtain the series $\sum_{n=1}^{\infty} -\frac{1}{n2^n}$ as the value for $\ln\left(1 + \frac{-1}{2}\right) = \ln\frac{1}{2} = -\ln 2$, so that $\ln 2 = \sum_{n=1}^{\infty} \frac{1}{n2^n}$. Since $-\frac{1}{2}$ is midway between the center and the endpoint of the interval of convergence, the rate of convergence will now be more rapid. Unfortunately, the series $\sum \frac{1}{n2^n}$ does not provide upper bounds for $\ln 2$.

Using $n = 10$, the first series $\sum_{n=1}^{\infty} \frac{(-1)^{n+1}}{n}$ gives the approximation $.645634 < \ln 2 < .745634$ with an error less than $\frac{1}{10}$, while the second series $\sum_{n=1}^{\infty} \frac{1}{n2^n}$ gives the approximation $.6930649 < \ln 2$. Because we know that $\ln 2 \approx .6931472$, we see that the second approximation is more accurate. But if we did not already know this value, the second series would not give a useful error estimate. $\square$

Example 9   Obtain the series for $\cos x$ from the series for $\sin x$.

Solution.   One approach is to write

$$\cos x = \frac{d}{dx}(\sin x) = \frac{d}{dx}\left(x - \frac{x^3}{3!} + \frac{x^5}{5!} - \frac{x^7}{7!} + - \cdots\right)$$

$$= 1 - \frac{3x^2}{3!} + \frac{5x^4}{5!} - \frac{7x^6}{7!} + - \cdots$$

$$= 1 - \frac{x^2}{2!} + \frac{x^4}{4!} - \frac{x^6}{6!} + - \cdots$$

Using an alternate approach,

$$- \cos x = \int \sin x \, dx$$

$$= \int \left(x - \frac{x^3}{3!} + \frac{x^5}{5!} - \frac{x^7}{7!} + - \cdots \right) dx$$

$$= C + \frac{x^2}{2} - \frac{x^4}{4 \cdot 3!} + \frac{x^6}{6 \cdot 5!} - \frac{x^8}{8 \cdot 7!} + - \cdots$$

Let $x = 0$ to show that $C = -1$ and therefore

$$\cos x = 1 - \frac{x^2}{2!} + \frac{x^4}{4!} - \frac{x^6}{6!} + - \cdots$$

We are able to differentiate and integrate term-by-term as we did be-cause there is uniform convergence on all closed subintervals of the interval of convergence $(-\infty, \infty)$ for $\sin x$.

EXERCISES 35

1. Find the set of pointwise convergence for the following series.
   (a) $\sum \frac{2^n (x-1)^n}{n}$
   (b) $\sum \frac{n+1}{3^n} x^{2n}$
   (c) $\sum \frac{e^{nx}}{2^n}$

2. What is the interval of convergence for the Maclaurin series expansion of $f(x) = \frac{1}{1-2x}$? It shortens the work to use geometric series results.

3. Given the Maclaurin series $\cos x = 1 - \frac{x^2}{2!} + \frac{x^4}{4!} - + \cdots$, find a value for $\int_0^1 \frac{1 - \cos x}{x^2} \, dx$ with an error less than 10^{-5}.

4. Find the Taylor series expansion for $f(x) = \sqrt{x}$ in powers of $(x - 9)$.

5. (a) Given the binomial expansion

$$(1+x)^r = 1 + rx + \frac{r(r-1)}{2!}x^2 + \frac{r(r-1)(r-2)}{3!}x^3 + \cdots$$

use Theorems 34.1–34.3 to find the Maclaurin series for arcsin x.

(b) Use your answer in part (a) to approximate π. Compare the use of $x = 1$ with the use of $x = \frac{1}{2}$.

6. Let $r = \frac{1}{2}$ in the series expansion for $(1+x)^r$. Use this to find an approximation for $\sqrt{3}$. Remember that the interval of convergence for the binomial series is $(-1, 1)$, so you cannot use $x = 2$.

7. (a) Show how to obtain the Maclaurin series

$$\arctan x = x - \frac{x^3}{3} + \frac{x^5}{5} - \frac{x^7}{7} + - \cdots$$

(b) Use your answer in part a to find a series approximation for π. How many terms are needed to find a value with an error less than .001?

(c) Use trig identities for $\tan(\alpha + \beta)$ and $\tan(4\alpha)$ to derive the formula

$$\frac{\pi}{4} = 4 \arctan \frac{1}{5} - \arctan \frac{1}{239}$$

(d) How many terms are needed to approximate π to 5 decimal place accuracy using the formula in part c?

(e) Use DeMoivre's Theorem and the expansion of $(5 - i)^4 (1 + i)$ in order to derive the same formula as in part c.

8. Find the value for each of the following series. Use the list of Maclaurin series as given in this section.

(a) $1 + \frac{1}{2} + \frac{1}{4} + \frac{1}{8} + \frac{1}{16} + \cdots$

(b) $1 - \frac{1}{2} + \frac{1}{3} - \frac{1}{4} + \frac{1}{5} - \frac{1}{6} + - \cdots$

(c) $1 - \frac{1}{3} + \frac{1}{5} - \frac{1}{7} + \frac{1}{9} - + \cdots$

(d) $1 + \frac{1}{3} + \frac{1}{5} + \frac{1}{7} + \cdots$

(e) $1 + \frac{1}{2} + \frac{1}{6} + \frac{1}{24} + \frac{1}{120} + \cdots$

(f) $1 + \frac{1}{4} + \frac{1}{9} + \frac{1}{16} + \frac{1}{25} + \cdots$

(g) $1 - \frac{1}{6} + \frac{1}{120} - \frac{1}{5040} + \cdots$

(h) $1 + \frac{1}{6} + \frac{1}{120} + \frac{1}{5040} + \cdots$

(i) $2 - \frac{2}{3} + \frac{2}{9} - \frac{2}{27} + \frac{2}{81} - + \cdots$

(j) $1 + \frac{1}{2} - \frac{1}{8} + \frac{1}{16} - \frac{5}{128} + - \cdots$

9. Prove Theorem 35.1.

SECTION 36   FOURIER SERIES

A section on Fourier series should be included in an analysis text for several reasons. One reason is that series of trigonometric functions provide a second large category of functions, in addition to series of polynomials, to illustrate the properties of uniform convergence. The results on Taylor series may be better understood through this comparison. A second reason is that Fourier series have played a very significant role in the historical development of mathematics. At the close of the eighteenth century, when most mathematicians believed that there were no new results left to be discovered in mathematics, the use of trigonometric series by Fourier to study the properties of heat opened new vistas for mathematicians to pursue, and the pace of discovery has not slackened since.

In addition, the new kinds of functions that arose through the study of trigonometric series expanded the application of calculus to many new types of functions, since previous work had almost entirely concentrated on continuous functions. Also, when Georg Cantor worked with Fourier series around 1870, he found that their sets of convergence involved many new kinds of sets, and he was motivated to initiate his fruitful study of the theory of infinite sets as a result. The implications of this one fact alone secure the place of Fourier series as an important part of mathematics.

In his study of the transfer of heat in solids and liquids at the start of the 19th century, Joseph Fourier was led to ask the following interesting mathematical question. Suppose f is a function that is defined on the interval $[-\pi, \pi]$. Is it possible to find constants $A_0, A_1, A_2, \ldots$ and $B_1, B_2, \ldots$ so that the trigonometric series

$$\sum_{n=0}^{\infty} A_n \cos nx + \sum_{n=1}^{\infty} B_n \sin nx$$

converges to $f(x)$ for all $x \in [-\pi, \pi]$? This question is very similar to asking whether a function defined on $[-r, r]$ has a power series that converges to it. This is not generally true at all, because such a function f must have derivatives of all orders at each $x \in [-r, r]$. Fourier and later researchers found that while not every function could be represented by a trigonometric series, there are far less restrictions on the function f in this situation.

A natural question to ask is how to determine the constants A_0, A_1, A_2, ... and B_1, B_2, ... We assume for now that there is uniform convergence on $[-\pi, \pi]$, so that we can multiply through by $\cos kx\, dx$ and then integrate term-by-term to obtain

$$\int_{-\pi}^{\pi} f(x)\cos kx\, dx \;=\; \sum_{n=0}^{\infty} \int_{-\pi}^{\pi} A_n \cos nx \cos kx\, dx$$

$$+ \sum_{n=1}^{\infty} \int_{-\pi}^{\pi} B_n \sin nx \cos kx\, dx$$

By the results in Exercise 27.4, this simplifies to

$$\int_{-\pi}^{\pi} f(x)\cos kx\, dx \;=\; \pi A_k$$

so that
$$A_k \;=\; \frac{1}{\pi}\int_{-\pi}^{\pi} f(x)\cos kx\, dx \qquad \text{if } k \geq 1$$

In a similar way, we can show that

$$A_0 \;=\; \frac{1}{2\pi}\int_{-\pi}^{\pi} f(x)\, dx$$

$$B_k \;=\; \frac{1}{\pi}\int_{-\pi}^{\pi} f(x)\sin kx\, dx \qquad \text{if } k \geq 1$$

This process should remind you of the situation with Maclaurin series in Sections 21 and 35 where we assumed that a function f could be represented by a power series of the form $f(x) = \sum_{n=1}^{\infty} a_n x^n$, and then showed that the coefficients could be expressed in terms of derivatives involving f, namely $a_n = \frac{f^{(n)}(0)}{n!}$. In the case of Fourier series, we see that the coefficients can be represented in terms of integrals involving $f(x)$.

Example 1 Find the Fourier series expansion for the function

$$f(x) \;=\; \begin{cases} 1 & \text{if } 0 < x \leq \pi \\ 0 & \text{if } \quad x = 0 \\ -1 & \text{if } -\pi \leq x < 0 \end{cases}$$

Solution. Since $f(x)$ is an odd function, $f(x)\cos nx$ is also odd, and so by Exercise 27.10

$$A_0 \;=\; \frac{1}{2\pi}\int_{-\pi}^{\pi} f(x)\,dx = 0$$

$$A_n \;=\; \frac{1}{\pi}\int_{-\pi}^{\pi} f(x)\cos nx\,dx = 0 \qquad \text{for all } n \ge 1$$

In addition,

$$
\begin{aligned}
B_n \;&=\; \frac{1}{\pi}\int_{-\pi}^{\pi} f(x)\sin nx\,dx \;=\; \frac{2}{\pi}\int_{0}^{\pi} f(x)\sin nx\,dx \\[2mm]
&=\; \frac{2}{\pi}\int_{0}^{\pi}\sin nx\,dx \;=\; \frac{2}{\pi}\left(\frac{-\cos nx}{n}\right)\Big|_{0}^{\pi} \\[2mm]
&=\; \frac{2}{\pi n}(-\cos n\pi + \cos 0) \;=\; \frac{2}{\pi n}\left(1-(-1)^n\right) \\[2mm]
&=\; \begin{cases} \dfrac{4}{\pi n} & \text{if } n \text{ is odd} \\[2mm] 0 & \text{if } n \text{ is even} \end{cases}
\end{aligned}
$$

Therefore the Fourier series for f is given by

$$
\begin{aligned}
f(x) \;&=\; \sum_{n \text{ odd}} \frac{4}{\pi n}\sin nx \\[2mm]
&=\; \frac{4}{\pi}\left(\sin x + \frac{\sin 3x}{3} + \frac{\sin 5x}{5} + \cdots\right)
\end{aligned}
$$

Notice that the series for f converges to 0 at $x = \pi$, but $f(\pi) = 1$. $\square$

Example 2   Use the Fourier series in Example 1 to show that

$$\frac{\pi}{4} \;=\; 1 - \frac{1}{3} + \frac{1}{5} - \frac{1}{7} + - \cdots$$

Solution.   By setting $x = \frac{\pi}{2}$ in the Fourier series for $f(x)$ in Example 1, we obtain the formula

$$
\begin{aligned}
1 \;&=\; \frac{4}{\pi}\left(\sin \frac{\pi}{2} + \frac{1}{3}\sin\frac{3\pi}{2} + \frac{1}{5}\sin\frac{5\pi}{2} + \cdots\right) \\[2mm]
&=\; \frac{4}{\pi}\left(1 - \frac{1}{3} + \frac{1}{5} - \frac{1}{7} + - \cdots\right)
\end{aligned}
$$

from which we obtain the desired equation. You may recognize this as the same formula that we obtained earlier in Section 35, by setting $x = 1$ in the Maclaurin series for $\arctan x$. □

Example 3 Find the Fourier series expansion for $f(x) = x^2$.

Solution. Since $x^2 \sin nx$ is an odd function, we know that for all n,

$$B_n = \frac{1}{\pi} \int_{-\pi}^{\pi} x^2 \sin nx \, dx = 0$$

Using Exercise 27.10 and integration by parts, we obtain

$$A_n = \frac{1}{\pi} \int_{-\pi}^{\pi} x^2 \cos nx \, dx = \frac{2}{\pi} \int_0^{\pi} x^2 \cos nx \, dx$$

$$= \frac{2}{\pi} \left(\frac{x^2 \sin nx}{n} + \frac{2x \cos nx}{n^2} - \frac{2 \sin nx}{n^3} \right) \Big|_0^{\pi}$$

$$= \frac{2}{\pi} \left(\frac{2\pi \cos n\pi}{n^2} \right) = \frac{4}{n^2}(-1)^n \quad \text{if } n \geq 1$$

In addition,

$$A_0 = \frac{1}{2\pi} \int_{-\pi}^{\pi} x^2 \, dx = \frac{\pi^2}{3}$$

so we have for the Fourier series expansion of x^2,

$$\frac{\pi^2}{3} + \sum_{1}^{\infty} (-1)^n \frac{4 \cos nx}{n^2} = \frac{\pi^2}{3} - 4\cos x + \frac{4 \cos 2x}{4} - \frac{4 \cos 3x}{9} + - \cdots \quad □$$

Example 4 Use the Fourier series for x^2 in Example 3 to show that $\sum_{n=1}^{\infty} \frac{1}{n^2} = \frac{\pi^2}{6}$. See also Example 2.8.

Solution. If we replace x by π in the last line of Example 3, since $\cos n\pi = (-1)^n$, we have,

$$\pi^2 = \frac{\pi^2}{3} + \sum_{n=1}^{\infty} \frac{4}{n^2}$$

from which it follows that

$$\sum_{n=1}^{\infty} \frac{1}{n^2} = \frac{\pi^2}{6} \quad □$$

Example 5   Find the Fourier series expansion for the function $f(x) = x^3$ on the interval $[-\pi, \pi]$.

Solution.   Since the functions x^3 and $x^3 \cos nx$ are odd functions, we know that

$$A_0 \;=\; \frac{1}{2\pi} \int_{-\pi}^{\pi} x^3 \, dx \;=\; 0$$

$$A_n \;=\; \frac{1}{\pi} \int_{-\pi}^{\pi} x^3 \cos nx \, dx \;=\; 0 \qquad \text{for } n \geq 1$$

Using integration by parts and a result from Example 3, we show that

$$\int_0^{\pi} x^3 \sin nx \, dx \;=\; \frac{-x^3 \cos nx}{n} \Big|_0^{\pi} + \frac{3}{n} \int_0^{\pi} x^2 \cos nx \, dx$$

$$=\; \frac{(-1)^{n+1}\pi^3}{n} + \frac{3}{n} \frac{(-1)^n 2\pi}{n^2}$$

$$=\; (-1)^n \frac{\pi}{n} \left(\frac{6}{n^2} - \pi^2 \right)$$

Therefore it follows that

$$B_n \;=\; \frac{1}{\pi} \int_{-\pi}^{\pi} x^3 \sin nx \, dx$$

$$=\; \frac{2}{\pi} \int_0^{\pi} x^3 \sin nx \, dx$$

$$=\; (-1)^n \frac{2}{n} \left(\frac{6}{n^2} - \pi^2 \right)$$

and the Fourier series for $f(x) = x^3$ is given by

$$\sum_{n=1}^{\infty} B_n \sin nx \;=\; \sum_{n=1}^{\infty} (-1)^n \frac{2}{n} \left(\frac{6}{n^2} - \pi^2 \right) \sin nx \qquad \square$$

One might be tempted to try to use the above Fourier series to evaluate $\sum_{n=1}^{\infty} \frac{1}{n^3}$, as we used the Fourier series for x^2 to evaluate $\sum_{n=1}^{\infty} \frac{1}{n^2}$ in Example 4. All attempts to select an appropriate value for x seem to fail however. For example, if we let $x = \frac{\pi}{2}$, since $\sum_{n=1}^{\infty} \sin \frac{n\pi}{2} = 1 + 0 - 1 + 0 + \cdots$, we obtain

$$\left(\frac{\pi}{2}\right)^3 = \sum_{n=1}^{\infty} \frac{(-1)^n 2}{n} \left(\frac{6}{n^2} - \pi^2\right) \sin \frac{n\pi}{2}$$

$$= 2\pi^2 \left(1 - \frac{1}{3} + \frac{1}{5} - \frac{1}{7} + - \cdots\right)$$

$$-12 \left(1 - \frac{1}{3^3} + \frac{1}{5^3} - \frac{1}{7^3} + - \cdots\right)$$

Since $1 - \frac{1}{3} + \frac{1}{5} - \frac{1}{7} + \cdots = \frac{\pi}{4}$, we have

$$\sum_{n=1}^{\infty} \frac{(-1)^{n+1}}{(2n-1)^3} = 1 - \frac{1}{3^3} + \frac{1}{5^3} - \frac{1}{7^3} + - \cdots = \frac{\pi^3}{32}$$

but this gives no information about the value for $\sum_{n=1}^{\infty} \frac{1}{n^3}$. □

Historical Comment. Euler established closed formulas for the series $\sum_{n=1}^{\infty} \frac{1}{n^{2k}}$ for $k = 1, 2, \ldots, 13$, showing that

$$\sum_{n=1}^{\infty} \frac{1}{n^{2k}} = r_k \pi^{2k}$$

where r_k is a rational number. The same results can be established by the use of Fourier series expansions of the functions $x^2, x^4, \ldots, x^{26}$. Though it might be reasonable to conjecture that $\sum_{n=1}^{\infty} \frac{1}{n^3}$ has the value of a rational multiple of π^3, no one has been able to prove or disprove this conjecture. It is now known that

$$\sum_{n=1}^{\infty} \frac{1}{n^{2k}} = \frac{(2\pi)^{2k}}{2\,(2k)!} |B_{2k}|$$

holds for all natural numbers k, where B_{2k} represents a Bernoulli number.

When we consider the Fourier series of a function such as $f(x) = x^2$ on the interval $[-\pi, \pi]$, since the trigonometric functions are periodic with period 2π, we assume that the represented function is also periodic with period 2π. Thus we are thinking of the function $f(x) = x^2$ as graphed in Figure 36.1.

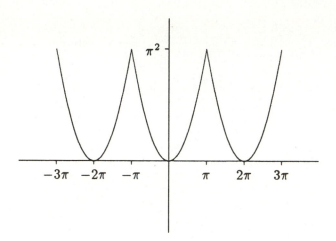

Figure 36.1

Clearly such a function is not differentiable at points such as $x = \pm\pi$ or $x = \pm 3\pi$. However, when we consider $f(x)$ on the interval $[-\pi, \pi]$, we can assume one-sided derivatives at the endpoints. For example,

$$f'(\pi) = \lim_{x \to \pi^-} f'(x) = 2\pi$$

We conclude this section and the book with a theorem that asserts the Fourier series of a function satisfying certain conditions converges uniformly on a certain set. This is by no means the most general result or proof possible, but it does illustrate the use of the Weierstrass M-Test.

THEOREM 36.1   Assume that the following conditions hold.

(1) The functions f, f', and f'' exists and are integrable on $[-\pi, \pi]$.
(2) $f(\pi) = f(-\pi)$
(3) f'' is bounded on $[-\pi, \pi]$; suppose $|f''(x)| \le K$ for $x \in [-\pi, \pi]$.

Let $B_n = \frac{1}{\pi} \int_{-\pi}^{\pi} f(x) \sin nx \, dx$. Then the series $\sum_{n=1}^{\infty} B_n \sin nx$ converges uniformly on $[-\pi, \pi]$.

Proof.   Using integration by parts twice (see Exercise 27.7), and the hypothesis that $f(-\pi) = f(\pi)$, we have

$$B_n = \frac{1}{\pi} \int_{-\pi}^{\pi} f(x) \sin nx \, dx$$

$$= -\frac{1}{\pi} \frac{f(x) \cos nx}{n} \Big|_{-\pi}^{\pi} + \frac{1}{\pi} \frac{f'(x) \sin nx}{n^2} \Big|_{-\pi}^{\pi}$$

$$-\frac{1}{\pi} \int_{-\pi}^{\pi} \frac{f''(x) \sin nx}{n^2} \, dx$$

$$= -\frac{1}{\pi} \frac{f(\pi)(-1)^n - f(-\pi)(-1)^n}{n} + \frac{1}{\pi} \frac{f(\pi) \cdot 0 - f(-\pi) \cdot 0}{n^2}$$

$$-\frac{1}{\pi} \int_{-\pi}^{\pi} \frac{f''(x) \sin nx}{n^2} \, dx$$

$$= -\frac{1}{\pi} \int_{-\pi}^{\pi} \frac{f''(x) \sin nx}{n^2} \, dx$$

and so

$$| B_n \sin nx | \leq | B_n | = \left| -\frac{1}{\pi} \int_{-\pi}^{\pi} \frac{f''(x) \sin nx}{n^2} \, dx \right|$$

$$\leq \frac{1}{\pi} \int_{-\pi}^{\pi} \frac{|f''(x)|}{n^2} \, dx$$

$$\leq \frac{B}{\pi} \cdot \frac{2\pi}{n^2} = \frac{2B}{n^2}$$

Since $\sum_{n=1}^{\infty} \frac{2B}{n^2}$ converges, $\sum_{n=1}^{\infty} B_n \sin nx$ converges uniformly on $[-\pi, \pi]$ by the Weierstrass M-Test.   ∎

Final Comment.   In spite of the hard work and trauma associated with the study of analysis in this book, I hope that you have enjoyed and profited from your efforts, and that you will continue your study of this fascinating and useful subject.

EXERCISES 36

1. Verify that $B_k = \frac{1}{\pi} \int_{-\pi}^{\pi} f(x) \sin kx \, dx$ for $k \geq 1$.

2. Prove that the Fourier series expansion for $f(x) = |x|$ is

$$\frac{\pi}{2} - \frac{4}{\pi} \left(\frac{\cos x}{1^2} + \frac{\cos 3x}{3^2} + \frac{\cos 5x}{5^2} + \cdots \right)$$

3. Use the formula in Exercise 2 to find a value for the series

$$1 + \frac{1}{3^2} + \frac{1}{5^2} + \cdots$$

4. Use the result in Example 4 that $\sum_{n=1}^{\infty} \frac{1}{n^2} = \frac{\pi^2}{6}$ in order to solve Exercise 3 by another method.

5. Prove that the Fourier series expansion for $f(x) = x$ is

$$x = \sum_{n=1}^{\infty} (-1)^{n+1} \frac{2}{n} \sin nx$$

6. Use the formula in Exercise 5 and an appropriate value for x in order to find the value for a convergent series of constants.

7. Furnish the verification in Example 5 that

$$\int_0^{\pi} x^3 \sin nx \, dx = (-1)^n \frac{\pi}{n} \left(\frac{6}{n^2} - \pi^2 \right)$$

8. Prove that the Fourier series expansion for $f(x) = x^4$ on the set $[-\pi, \pi]$ is

$$x^4 = \frac{\pi^4}{5} + 8 \sum_{n=1}^{\infty} \left(\frac{\pi^2}{n^2} - \frac{6}{n^4} \right) (-1)^n \cos nx$$

9. Use the result in Exercise 8 to show that $\sum_{n=1}^{\infty} \frac{1}{n^4} = \frac{\pi^4}{90}$.

10. Prove that the Fourier series expansion for $f(x) = e^x$ is

$$\frac{\sinh \pi}{\pi} + \frac{2 \sinh \pi}{\pi} \sum_{n=1}^{\infty} (-1)^n \frac{\cos nx}{1 + n^2} + \frac{2 \sinh \pi}{\pi} \sum_{n=1}^{\infty} (-1)^{n+1} \frac{n \sin nx}{1 + n^2}$$

11. Establish the integral formula

$$\int_0^{\pi} x^{2t} \cos nx \, dx = (-1)^n \sum_{k=1}^{t} (-1)^{k+1} \frac{(2t)! \, \pi^{2t+1-2k}}{(2t + 1 - 2k)! \, n^{2k}}$$

12. Use mathematical induction to establish the formula that

$$\sin^{2n} x = \frac{1}{2^{2n}}\binom{2n}{n} + \sum_{i=1}^{n} \frac{(-1)^i}{2^{2n-1}}\binom{2n}{n-i} \cos 2ix$$

This shows that $\sin^{2n} x$ has a finite Fourier series representation.

13. Use Exercise 12 to show that $\int_0^\pi \sin^{2n} x \, dx = \frac{1}{2^{2n}}\binom{2n}{n}\pi$.

14. Use Exercise 12 to show that

$$\sin^6 x = \frac{5}{16} - \frac{15}{32}\cos 2x + \frac{3}{16}\cos 4x - \frac{1}{32}\cos 6x$$

15. Use DeMoivre's formula and the binomial theorem to obtain a formula for $\sin^{2n-1} x$, similar to the one obtained in Exercise 12 for $\sin^{2n} x$.

16. Find where a problem arises when the steps in Theorem 36.1 are applied to the series $\sum_{n=1}^{\infty} A_n \cos nx$.

Appendix 1
Sets

I assume that you have been exposed to the basic ideas of set theory, and therefore I can simply list the usual symbols and definitions here as a reference. A set is a collection of objects, or elements, and is usually denoted by a capital letter such as S, while a lower case letter such as s refers to an object in the set. All objects are considered to be part of some universal set $\mathcal{U}$. In this book, the universal set is usually the set of real numbers, which we denote by the letter $\mathbf{R}$.

The expression $s \in S$ will be read as "the element s belongs to the set S," while $s \notin S$ will mean that the element s does not belong to the set S. The collection of all objects in the universal set $\mathcal{U}$, but not in the set S, is called the *complement* set of S and is denoted by a symbol such as $\mathcal{C}S$ or $\mathcal{U}/S$. A special set is the empty set, which contains no elements and is denoted by ϕ. It follows that $\mathcal{C}\mathcal{U} = \phi$ and $\mathcal{C}\phi = \mathcal{U}$.

We are often interested in sets of objects that make up part of a given set. We say that T is a *proper subset* of S if every element in T is also in S, and there is at least one element in S that is not in T. This is expressed by $T \subset S$, while the symbol $T \subseteq S$ means either that $T \subset S$ or $T = S$. Some of the well-known subsets of the set of real numbers are the set $\mathbf{N}$ of natural numbers, the set $\mathbf{Z}$ of integers, the set $\mathbf{Q}$ of rational numbers, and the set $\mathbf{P}$ of irrational numbers. We can write $\phi \subset \mathbf{N} \subset \mathbf{Z} \subset \mathbf{Q} \subset \mathbf{R}$ to illustrate the subset idea. It is also true that $\mathbf{P} = \mathcal{C}\mathbf{Q} = \mathbf{R}/\mathbf{Q}$.

We can represent a set in different ways. One way that works well for sets with few elements is to simply enumerate the elements within brackets, such as $A = \{2, 3, 5, 7\}$. Another way is to use the set-builder notation; for example, $B = \{x \in \mathbf{N} : x \leq 10 \text{ and } x \text{ is prime}\}$ is read as "B is the set of all natural numbers so that x is less than or equal to 10 and x is prime." You may observe that the sets A and B are identical, although expressed in different ways.

Perhaps the most familiar type of set that we encounter in our study of analysis is the interval. The open interval (a, b) is defined as the set $\{x \in \mathbf{R} : a < x < b\}$, while the closed interval $[a, b] = \{x \in \mathbf{R} : a \leq x \leq b\}$. There are also the half-open intervals $[a, b) = \{x \in \mathbf{R} : a \leq x < b\}$ and $(a, b] = \{x \in \mathbf{R} : a < x \leq b\}$. In addition, we have such unbounded intervals as $[a, \infty) = \{x \in \mathbf{R} : x \geq a\}$ or $(-\infty, b) = \{x \in \mathbf{R} : x < b\}$. The symbol of ∞ is no doubt also familiar to you. We can represent the set **R** of real numbers by the unbounded interval $(-\infty, +\infty)$.

Finally, there are operations defined on the collection of all the subsets of a universal set. If S and T are subsets of $\mathcal{U}$, we can form new sets by the following definitions.

The set of all elements that belong both to S and T is called the *intersection* of S and T, and is denoted by

$$S \cap T = \{x \in \mathcal{U} : x \in S \text{ and } x \in T\}.$$

The set of all elements belonging either to S or T, or possibly both, is called the *union* of S and T, and is denoted by

$$S \cup T = \{x \in \mathcal{U} : x \in S \text{ or } x \in T\}$$

As an example, $\mathbf{Q} \cup \mathbf{P} = \mathbf{R}$ and $\mathbf{Q} \cap \mathbf{P} = \phi$. In general,

$$\phi \subseteq S \cap T \subseteq S \subseteq S \cup T \subseteq \mathcal{U}.$$

An important part of set theory is concerned with the concept of the cardinal number of a set, which can be informally thought of as the number of elements in the set. The simplest distinction is between a finite set (one with finitely many elements) and an infinite set (one with an infinite number of elements). A more difficult idea is whether there is any distinction among the cardinal number of infinite sets. Prior to the time of Georg Cantor near the end of the nineteenth century, no such distinction was made by mathematicians. Then Cantor defined a set to be *countably infinite* if it could be placed in a 1-1 correspondence with the set **N** of natural numbers, and to be *uncountable* if it was infinite and could not be placed in such a correspondence. Thus there are at least two kinds of infinite sets (or levels of infinity) among the subsets of the real numbers. Cantor showed that the sets **Z** of the integers and **Q** of the rational numbers are countably infinite, while the sets **P** of irrational numbers and **R** of the real numbers are uncountable. The countable nature of the set of natural numbers is of importance for the definition of a sequence in Chapter 2.

We should also observe the similarities of and differences between the concepts of a sequence $\{x_n\}_{n=k}^{\infty}$ and the set $S = \{x_n : n \in \mathbf{N} \text{ and } n \geq k\}$ that is determined by the sequence $\{x_n\}_{n=k}^{\infty}$. One of the differences is that there is no ordering of elements or duplication of elements in a set. For example, the sequence $\{0, 1, 0, 1, \ldots\}$ has a countably infinite number of elements, while the set $S = \{0, 1\}$ determined by this sequence is finite. One similarity is that the concept of a limit point can be defined for both sequences and sets.

Appendix 2
The Structure of the Real Numbers

In high school, a mathematics student learns that the familiar operations of addition and multiplication in the set of real numbers satisfy properties such as the commutative and associative laws, and also the existence of identity and inverse elements. We can express these ideas in a more formal sense by the statement that the set $\mathbf{R}$ of real numbers is a *field* under the operations of addition $+$ and multiplication $\cdot$ because the following axioms are satisfied in $\mathbf{R}$.

Axiom 1 (Closure law)

If $x, y \in \mathbf{R}$, then $x + y \in \mathbf{R}$ and $x \cdot y \in \mathbf{R}$

Axiom 2 (Associative law)

If $x, y, z \in \mathbf{R}$, then $x + (y + z) = (x + y) + z$ and $x \cdot (y \cdot z) = (x \cdot y) \cdot z$

Axiom 3 (Commutative law)

If $x, y \in \mathbf{R}$, then $x + y = y + x$ and $x \cdot y = y \cdot x$

Axiom 4 (Distributive law)

If $x, y, z \in \mathbf{R}$, then $x \cdot (y + z) = x \cdot y + x \cdot z$

Axiom 5 (Identity elements)

There are unique elements $0, 1 \in \mathbf{R}$ so that $x + 0 = 0 + x = x$ and $x \cdot 1 = 1 \cdot x = x$ for all $x \in \mathbf{R}$

Axiom 6 (Inverse elements)

For every $x \in \mathbf{R}$, there is a unique element $-x \in \mathbf{R}$ so that $x + (-x) = (-x) + x = 0$, and for each $x \neq 0 \in \mathbf{R}$, there is a unique element $x^{-1} \in \mathbf{R}$ so that $x \cdot x^{-1} = x^{-1} \cdot x = 1$

Example 1 The set $\mathbf{N}$ of natural numbers is closed under the operations of $+$ and $\cdot$, and has a multiplicative identity of 1. No element has an additive inverse, and only the element 1 has a multiplicative inverse.

Example 2 In the set of real numbers, the additive inverse of 2 is -2, while the multiplicative inverse of 2 is $\frac{1}{2}$.

Example 3 The set **Z** of the integers has an additive identity of 0 and additive inverses of $-n$ for each integer n, but still no multiplicative inverses except for the elements of $+1$ and -1.

The set **Q** of rational numbers is also a field, although it lacks the property of completeness, which is discussed in Section 5. The set of complex numbers is an example of a field that contains the real number field as a proper subset.

The real number field is called an *ordered* field because there is an order relation $(<)$ defined on the elements in **R**, which satisfies the transitive law and the trichotomy law. By definition, $a < b$ (read a is less than b) means that $b - a$ is a positive number. Also $a \leq b$ means that either $a < b$ or $a = b$, while $a > b$ means that $b < a$. The following properties of the order relation in **R** are now listed for a reference.

Assume that $a, b, c \in \mathbf{R}$. Then

(1) If $a < b$ and $b < c$, then $a < c$ (Transitive law)

(2) Exactly one of the following holds:
$\qquad a < b, a = b,$ or $b < a$ (Trichotemy law)

(3) If $a < b$, then $a + c < b + c$

(4) If $a < b$ and $c > 0$, then $ac < bc$

(5) If $a < b$ and $c < 0$, then $ac > bc$

(6) If $0 < a < 1$, then $0 < a^2 < a < 1$

The complex number field does not have a familiar order relation defined on its elements, so this is one structural difference between the real number field and the complex number field. On the other hand, the field of rational numbers is an ordered field. We will see in Section 5 that the rational number field is not a *complete* ordered field, and it is at this place that the rational number field is structurally different from the real number field.

Appendix 3
Logic and Proof

If you have studied a unit on logic in a discrete mathematics course, you are familiar with the following notation. Let p and q represent two statements, which can be assigned either the value of true (T) or false (F). From these we form the following statements. The *negation* of p is denoted by $\neg p$, and has the opposite truth value of p. The *disjunction* $p \vee q$ is true whenever either p or q is true, and the *conjunction* $p \wedge q$ is true whenever p and q are both true.

The *conditional* $p \to q$ is true except when p is true and q is false. Many theorems of mathematics can be viewed as having this form, where p is the hypothesis of the theorem and q is the conclusion. In Lemma 4.2 for example, p is the statement that the sequence $\{x_n\}$ is convergent, while q is the statement that the sequence $\{x_n\}$ is bounded.

The statement $q \to p$ is called the *converse* of the statement $p \to q$. It is possible for $p \to q$ to be true and $q \to p$ to be false, as is the case in Lemma 4.2. The statement $p \leftrightarrow q$ is an abbreviation for $(p \to q) \wedge (q \to p)$, which is called a *biconditional*. In words, we then say that p is a necessary and sufficient condition for q, or that p is true if and only if q is true. This is sometimes shortened to p iff q.

The statement $\neg q \to \neg p$ is called the *contrapositive* of $p \to q$, and it is shown in logic that these are equivalent statements. This has significance for our proofs because if we prove either form, we have thereby proved the other. For example, once we have proved in Lemma 4.2 that all convergent sequences are bounded, we have also proved the contrapositive that all unbounded sequences diverge.

The contradictory statement $r \wedge \neg r$, which is always false, can never occur in a consistent system. Any statement that implies a contradiction such as $r \wedge \neg r$ must therefore be false. We next make some comments about different proof forms that are used in mathematics to prove theorems. You should watch for the occurrence of these forms throughout the book.

(1) Direct proof $[(p \to r) \wedge (r \to q)] \to (p \to q)$

We begin with the assumption that p is true and try to obtain that q is true directly from this. More often than not, we will first show that p implies some other statement r and then that r implies q. By the transitive property of $\rightarrow$, we conclude that $p \rightarrow q$. There could be many intermediate implications between p and q. The proof of Lemma 4.2 is a direct proof.

(2) Contrapositive proof $(\neg q \rightarrow \neg p) \leftrightarrow (p \rightarrow q)$

We first show by a direct proof that $\neg q \rightarrow \neg p$, and then use the fact that $p \rightarrow q$ and $\neg q \rightarrow \neg p$ are equivalent statements, so that if one is true, the other must also be true. Proofs of Lemma 5.1 and 5.2 are of this form.

(3) Proof by contradiction $[(p \wedge \neg q) \rightarrow (r \wedge \neg r)] \rightarrow (p \rightarrow q)$

This proof form is also called proof by denial or indirect proof. The method assumes both p and $\neg q$ and produces some contradictory statement $r \wedge \neg r$ as a direct implication. Since p is assumed to be true and we cannot have a contradiction, the only possible conclusion is that $\neg q$ cannot be true, so that q must be true. This proof form has been vigorously debated in the philosophy of mathematics during the past century, and has not been accepted as valid by all mathematicians. Since a large number of desirable theorems in mathematics are proved by this form of reasoning, the majority of mathematicians accept it as valid.

A good example of this type of proof is the one in Theorem 16.1, in which p is the statement that the function f is continuous on the closed interval $[a, b]$ and q is the statement that f is bounded on $[a, b]$. We begin this proof by assuming $\neg q$. If we could obtain $\neg p$ directly, then $p \rightarrow q$ would be true by the contrapositive argument. But if we must also assume that p is true in order to obtain a contradiction (as we do in this proof), then the proof by contradiction form is the one used.

(4) Converse statement $(q \rightarrow p) \rightarrow (p \rightarrow q)$

Although this form is often used by students, it is <u>not</u> a valid proof form. After proving a result of the form $p \rightarrow q$, we usually ask whether the converse is also true. More often than not, it will not be true and we demonstrate this by finding a counterexample, where q and $\neg p$ are both true. The analysis method of the Greeks, which was first widely used by Plato, was the attempt to prove $p \rightarrow q$ by first demonstrating the converse $q \rightarrow p$. If this could be done, Plato would say that it was more likely that $p \rightarrow q$ is true, but this could only be proved by reversing the steps in the argument for $q \rightarrow p$.

This method of working a proof backwards is really a method of proof discovery. Imre Lakatos developed this idea in much detail in his book *Proofs and Refutations*, published by Cambridge University Press in 1976. By looking for consequences of q, Lakatos says that if $q \rightarrow \neg p$ is shown, then we know it is impossible that $p \rightarrow q$ could hold, because this would yield $p \rightarrow \neg p$ by the transitive law. In this case, Lakatos urges the reader to find what part of the hypothesis q caused the contradiction and, after eliminating this problem part, to obtain a new hypothesis q'. The statement that $p \rightarrow q'$ should next be examined in a similar way.

(5) Law of Modus Ponens $[p \wedge (p \rightarrow q)] \rightarrow q$

This axiom of logic is contained in the *Principia Mathematica*, a 3 volume book on logic written in the early years of the twentieth century by Bertrand Russell and Alfred North Whitehead. This law states that if $p \rightarrow q$ is proven true, and p is known to be true, then q must also be true. Whenever we use a previously proved result as a reason in the proof of a theorem, we are using this axiom. A good illustration of this law is seen in Section 20.

Let Rolle's Theorem (Theorem 20.2) be written in the form $p \rightarrow q$, where p is the statement that the function F is continuous on the closed interval $[a, b]$, differentiable on the open interval (a, b), and such that $F(a) = F(b) = 0$, and q is the statement that there is some value $c \in (a, b)$ so that $F'(c) = 0$.

After we have proved Rolle's Theorem, we can assume that $p \rightarrow q$ is true. In order to prove the Mean Value Theorem (Theorem 20.3), we assume that the function f is continuous on $[a, b]$ and differentiable on (a, b), and form the new function $F(x) = f(x) - f(a) - M(x - a)$, where $M = \frac{f(b) - f(a)}{b - a}$. We then show that F meets the condition p in Rolle's Theorem, and therefore F must also satisfy the condition q by the Law of Modus Ponens. This yields the desired conclusion. Taylor's Theorem with Remainder (Theorem 21.1) is proved by this same technique.

(6) Principle of Mathematical Induction

This principle is another axiom, presented near the end of the nineteenth century by the Italian mathematician, Giuseppi Peano, as one of five axioms to define the set of natural numbers **N**. This proof method only applies to results that can be stated in the form of a statement $P(n)$, where n is a natural number. For example, we can use this method to prove that $1 + 2 + \cdots + n = \frac{n(n+1)}{2}$ for all $n \in$ **N**, because we can write this in the form $P(n) : 1 + 2 + \cdots + n = \frac{n(n+1)}{2}$. This inductive proof form is discussed and illustrated in Example 2.1, and used in the proofs of important results such as the Bolzano-Weierstrass Theorem for sets (Theorem 5.6) and the Heine-Borel Theorem (Theorem 13.6).

In addition to the above categories of proof according to the form of the logical arguments, there is another way to categorize some of the proofs encountered—namely as existence proofs, uniqueness proofs, and proofs by construction. An *existence theorem* refers to a result that asserts the existence of some quantity, without providing a means for actually finding this quantity. The Bolzano-Weierstrass Theorem (Theorem 7.1) is such a result, which asserts the existence of a convergent subsequence of a bounded sequence, but provides no method for finding it. By contrast, a *construction theorem* not only asserts the existence of a certain quantity, but also provides a method for finding it. The Product Rule for differentiation (Theorem 19.2) is such an example. Not only does this theorem tell us that the derivative $(f \cdot g)'$ exists if f' and g' exist, but also asserts that the value for $(f \cdot g)'$ is $f \cdot g' + f' \cdot g$.

Once we have proved that a certain quantity exists, we often ask next how many different quantities satisfy the conditions of the theorem. A *uniqueness theorem* proves that there is exactly one solution to the given problem. The fact that a convergent sequence has only one limit or that a non-empty bounded set has only one supremum are examples of uniqueness results.

Bibliography

Category 1: Analysis textbooks

Apostol, T. M. *Mathematical Analysis*, 2nd ed. Reading, Massachusetts: Addison-Wesley, 1974.

Bartle, Robert G., and Donald R. Sherbert. *Introduction to Real Analysis*. New York: John Wiley and Sons, 1982.

Buck, R. Creighton. *Advanced Calculus*, 2nd ed. New York: McGraw-Hill, 1965.

Fulks, Watson. *Advanced Calculus*, 2nd ed. New York: John Wiley and Sons, 1969.

Gaughan, Edward D. *Introduction to Analysis*, 3rd ed. Monterey, CA: Brooks/Cole Publishing, 1987.

Lay, Steven R. *Analysis–An Introduction to Proof.* Englewood Cliffs, NJ: Prentice-Hall, 1986.

Rudin, Walter. *Principles of Mathematical Analysis*, 2nd ed. New York: McGraw-Hill Book Company, 1964.

Taylor, Angus E., and W. Robert Mann. *Advanced Calculus*, 2nd ed. Lexington, Massachusetts: Xerox College Publishing, 1972.

Category 2: Historical materials (Books)

Baron, Margaret E. *The Origins of the Infinitesmal Calculus*. New York: Dover Publications, 1987.

Bell, E. T. *Men of Mathematics*. New York: Simon and Schuster, 1965; Seventh paperback printing of the 1937 edition.

Bottazzini, Umberto. *The Higher Calculus: A History of Real and Complex Analysis from Euler to Weierstrass*. New York: Springer-Verlag, 1986.

Boyer, Carl B. *The History of the Calculus and its Conceptual Development.* New York: Dover, 1959; paperback edition of the 1949 book *The Concepts of the Calculus.*

Boyer, Carl B., and Uta C. Merzbach. *A History of Mathematics*, 2nd ed. New York: John Wiley and Sons, 1989.

Burton, David M. *The History of Mathematics*. Boston: Allyn and Bacon, Inc., 1985.

Edwards, C. H. Jr. *The Historical Development of the Calculus*. New York: Springer-Verlag, 1979.

Eves, Howard. *An Introduction to the History of Mathematics*, 5th ed. Philadelphia: Saunders College Publishing, 1983.

Grabiner, Judith V. *The Origins of Cauchy's Rigorous Calculus*. Cambridge, Massachusetts: The MIT Press, 1981.

Grattan-Guinness, Ivor. *The Development of the Foundations of Mathematical Analysis from Euler to Riemann*. Cambridge, Massachusetts: The MIT Press, 1970.

Hofmann, Joseph E. *Leibniz in Paris 1672-1676*. Cambridge: Cambridge University Press, 1974.

Klambauer, Gabriel. *Aspects of Calculus*. New York: Springer-Verlag, 1986.

Kline, Morris. *Mathematical Thought from Ancient to Modern Times*. New York: Oxford University Press. 1980.

Lakatos, Imre. *Proofs and Refutation–The Logic of Mathematical Discovery*. Cambridge: Cambridge University Press, 1976.

Sondheimer, Ernst, and Alan Rogerson. *Numbers and Infinity*. Cambridge: Cambridge University Press, 1981.

Toeplitz, Otto. *The Calculus, a Genetic Approach*. Chicago: The University of Chicago Press, 1963.

Category 3:   Historical Materials (Articles)

Cajori, F. "History of the exponential and logarithmic concepts," *The American Mathematical Monthly*, **20**, pp. 5–14, 35–47, 1913.

Coolidge, J. L. "The Story of the Binomial Theorem," *The American Mathematical Monthly*, **56**, pp. 147–157, 1949.

Grabiner. Judith V. "Is Mathematical Truth Time–Dependent?," *The American Mathematical Monthly*, **81**, pp. 354–365, 1974.

Grattan-Guinness, Ivor. "Bolzano, Cauchy and the 'New Analysis' of the Early Nineteenth Century," *Archive for History of Exact Sciences*, **6**, pp. 372–400, 1969–1970.

Kleiner, Israel. "Evolution of the Function Concept: A Brief Survey," *The College Mathematics Journal*, **20**, No. 4, pp. 282–300, Sept 1989.

Langwitz, Detlef. "Infinitely Small Quantities in Cauchy's Textbook," *Historia Mathematica*, **14**, pp. 258–274, August, 1987.

Mancosu, Paolo. "The Metaphysics of the Calculus: A Foundational Debate in the Paris Academy of Sciences, 1700–1706," *Historia Mathematica*, **16**, No. 3, pp. 224–248, August 1989.

Monna, A. F. "The Concept of Functions in the 19th and 20th Centuries," *Archive for History of Exact Sciences*, **9**, pp. 57–83, 1972–1973.

Whiteside, D. T. "Patterns of Mathematical Thought in the later 17th Century," *Archive for History of Exact Sciences*, **1**, pp. 179–388, 1960–1962.

Youschkevitch, A. P. "The Concept of Function up to the Middle of the 19th Century," *Archive for History of Exact Sciences*, **16**, pp. 37–85, 1976.

Category 4:   Original Writings

Birkhoff, Garrett, ed. *A Source Book in Classical Analysis.* Boston: Harvard University Press, 1973.

Euler, L. *Introduction to Analysis of the Infinite–Book 1.* Translated by John D. Blanton. New York: Springer-Verlag, 1988.

Smith, David Eugene, ed. *A Source Book in Mathematics.* New York: Dover Publications, 1959.

Struik, Dirk, ed. *A Source Book in Mathematics, 1200–1800.* Cambridge, Massachusetts: Harvard University Press. 1969.

Answers for Selected Exercises

EXERCISES 4 (page 37)

1. (a) 0   Hint: Consider the rationalizing factor $\sqrt{n+1} + \sqrt{n}$.
   (b) $\frac{1}{2}$
2. $x_5 = \frac{6}{10}$ $\lim x_n = \frac{1}{2}$
3. (a) $N = \left[\sqrt{(\frac{1}{\epsilon} - 1)}\right]$ (c) $N = \left[\frac{9-5\epsilon}{2\epsilon}\right]$
4. (b) $N(.01) = 13$
5. (a) Hint: See Exercise 17b.   (b) $x_n = (-1)^n$
6. (b) No
7. Let $x_n = \cos n$ and $y_n = \sec n$.
12. Hint: See Example 5 and use the identity that
    $a^3 - b^3 = (a-b)(a^2 + ab + b^2)$.
13. One way is to show that $|x_n y_n z_n - xyz| < \epsilon$. A shorter way is
    to use Result (2) from Theorem 4.1.
15. $+\infty$ if $c > 1$,   1 if $c = 1$,   0 if $|c| < 1$,   does not exist if $c \leq -1$

EXERCISES 5 (page 47)

1. (a) $\sup S = +\infty$ $\inf S = -\infty$
   (b) $\sup S = \frac{5}{2}$ $\inf S = 1$
   (d) $\sup S = \sqrt{2}$ $\inf S = 0$
2. If a non-empty set has a lower bound, it must have a greatest lower
   bound.
11. Hint: Consider the set $-S$, as defined in Exercise 10.
12. (a) and (d)
13. (a) 0   (b) None   (c) all real numbers in $[0,1]$
16. $\cap I_n$ is empty in Example 5, but $\cap I_n = \{1\}$ in Example 6.
17. You must prove that the Bolzano-Weierstrass Theorem implies the
    LUB axiom.
18. S is not known to be bounded, so Theorem 5.6 does not apply.

EXERCISES 6 (page 55)

2. $\{x_n\}$ is nonincreasing for $n \geq 2$
3. $x_8 = \frac{107}{64}$
4. (b) Observe that $\sum_{n=2}^{k} \left(\frac{1}{n-1} - \frac{1}{n}\right) = 1 - \frac{1}{k}$
5. (b) $\lim x_n = 2$
6. Diverges

7. Show that $\{x_n\}$ is increasing and bounded above by 1.
12. $a^7 + 7a^6b + 21a^5b^2 + 35a^4b^3 + 35a^3b^4 + 21a^2b^5 + 7ab^6 + b^7$

EXERCISES 7 (page 61)

1. (a) 0, ± 1 (b) -1, 0   (c) None
2. (a) None   (d) $\frac{1}{2}$
4. $\frac{a+2b}{3}$
8. (a) $\sqrt[3]{e}$
   (b) Show first that $1 - \frac{1}{n} = \frac{1}{1 + \frac{1}{n-1}}$.
   (c) $+\infty$
   (d) Show that $\{x_{3n}\}$ converges to e^3. What does this imply about $\{x_n\}$?
9. (a) $\frac{1}{3}$ (b) 0   (c) 0

EXERCISES 8 (page 67)

1. For the product case, find N so that $|x_n y_n - x_m y_m| < \epsilon$, for $n, m > N$.
2. (a) $|x_n - x_m| = \frac{2|n-m|}{nm} < \frac{2}{m}$
3. $|x_{m+1} + \cdots + x_n| < \epsilon$, if we assume $m < n$

EXERCISES 9 (page 73)

1. (a) $\frac{3}{2}$ (c) 5
2. Consider the nth term test.
4. Using Method 1, show that $1 + \frac{1}{2} + \cdots + \frac{1}{n} > \int_1^{n+1} \frac{1}{x}\, dx$.
5. Hint: Observe that the sequence $\{S_n\}$ is increasing, and use Theorem 6.1.
7. (a) and (b) converge   (e) and (f) diverge

EXERCISES 10 (page 80)

1. (a) Let $\delta(\epsilon) = \min\{1, \frac{\epsilon}{6}\}$ or $\min\{2, \frac{\epsilon}{7}\}$ or $\min\{\frac{1}{2}, \frac{2\epsilon}{7}\}$
   (d) Let $\delta(\epsilon) = \min\{1, \frac{40\epsilon}{3}\}$
   (e) Let $\delta(\epsilon) = \min\{\frac{1}{2}, \frac{45\epsilon}{26}\}$
   (f) Let $\delta(\epsilon) = \min\{1, (2 + \sqrt{3})\epsilon\}$ or $\delta(\epsilon) = 2\epsilon$
3. Find $\delta(\epsilon)$ so that $|f(x)g(x) - LM| < \epsilon$, for all x so that $|x - c| < \delta(\epsilon)$

EXERCISES 11 (page 87)

3. (a) Let $\delta(R) = \min\{1, \frac{3}{R}\}$

(b) Let $D(\epsilon) = \sqrt[3]{\frac{2}{\epsilon}}$

(c) Let $\delta(R) = \sqrt[3]{\frac{2}{R}}$

(d) Let $\delta(\epsilon) = \min\{\frac{|c|}{2}, \frac{\epsilon|c|^2}{2}\}$.

EXERCISES 12 (page 91)

1. (b) Let $y_n = \frac{2}{\pi}\left(\frac{1}{1+4n}\right)$

EXERCISES 13 (page 96)

4. Let $I_n = (a - \frac{1}{n}, a + n)$

7. Let $T_n = \mathbf{R}\backslash S_n$ and show that $T_{n+1} \subset T_n$.

EXERCISES 15 (page 104)

1. (a) $x = -2$ is removable, $x = 2$ is infinite

(b) Jump discontinuities at $x = n + \frac{1}{2}$, for $n \in \mathbf{Z}$.

(e) Infinite discontinuities at $x = n\pi$, for $n \in \mathbf{Z}$

(h) See Example 2.2

(k) Jump discontinuity at $x = 0$

(m) Infinite discontinuities at $x = 0$ and $x = 1$

2. (b) $\delta = \min\{\frac{|c+2|}{2}, \frac{\epsilon|c+2|^2}{2}\}$

(c) $\delta = \min\{1, \frac{\epsilon}{3c^2+3|c|+1}\}$

7. Use mathematical induction

8. (a) Let $\{x_n\} = \{2 + \frac{1}{n}\}$.

EXERCISES 16 (page 111)

1. $m = -5,\quad M = \frac{17}{3}$

2. One zero is between 1 and 2; the other is between -8 and -7.

3. One possibility is $f(x) = \frac{1}{x}$ if $0 < x \le 1$, and $f(0) = 0$.

5. All y so that $1 \le y \le 21$.

7. Consider the function $f(x) = x - \cos x$ on $[0, \frac{\pi}{2}]$.

EXERCISES 17 (page 115)

1. (a) Let $\delta(\epsilon) = \frac{\epsilon|c|^3}{10}$

(b) Let $\delta(\epsilon) = \frac{\epsilon}{3b^2}$

3. (b) Let $x_n = 1 + \frac{1}{n}$ and $u_n = 1 + \frac{1}{2n}$ or $u_n = 1 + \frac{2}{n}$
   (d) Try $x_n = n\pi$ and $u_n = n\pi + \frac{1}{n}$
5. For sets of the form $(-\infty, b]$ or $[b, \infty)$, show that $|f(x) - f(u)| =$
    $\frac{1}{1+x^2} \frac{1}{1+u^2} |x + u||x - u| \leq \left(\frac{1}{|u|} + \frac{1}{|x|} \right) |x - u|.$
7. (b) $f(x) = x^{\frac{2}{3}}$ is one possibility.

EXERCISES 19 (page 128)

1. (d) The rationalizing factor is $(x + h)^{\frac{2}{3}} + x^{\frac{1}{3}}(x + h)^{\frac{1}{3}} + x^{\frac{2}{3}}$
3. First use the definition of the derivative to show that
    $f'(x) = \frac{1}{x} \lim_{h \to 0} \log_b (1 + \frac{h}{x})^{\frac{x}{h}}$
6. See Example 2.2
7. $x_n = n(\sqrt[n]{b} - 1)$
9. (a) Use the fact that if $y = x^{\frac{p}{q}}$, then $y^q = x^p$.
   (f) If $y = \operatorname{arcsec} x$, then $x = \sec y$ and $\frac{dy}{dx} = \frac{1}{x\sqrt{x^2-1}}$. You must
        still decide whether to take the positive and negative square,
        root by considering the graph of $\operatorname{arcsec} x$.
10. (c) $\frac{1}{2\sqrt{x}(1+x)}$
13. (d) $f^{(n)}(1) = (-1)^{n+1} \frac{1 \cdot 3 \cdot 5 \cdot 7 \cdots (2n-3)}{2^n}$ for $n \in \mathbf{N}$
    (e) No simple pattern exists.
15. Observe that $[x + n] = [x] + n$, for $n \in \mathbf{N}$

EXERCISES 20 (page 137)

1. (b) $c = e - 1$
   (c) $c = \frac{\sqrt{4\pi^2 - 9}}{2\pi}$
2. (b) Decreasing on $(-1, \frac{1}{2})$ and $(\frac{1}{2}, 2)$
   (c) Increasing on $(0, \sqrt{e})$
3. (c) $|f'(x)| \leq \frac{3}{2}\sqrt{c}$ on the interval $[0, c]$
4. (a) On sets of the form $(-\infty, d]$ where $d < -1$ or $[c, \infty)$
        where $c > -1$
5. (b) $3\frac{1}{12} < \sqrt[3]{31} < 3\frac{4}{27}$
   (d) $\frac{7+\pi}{4} < \sin 1.5 < 1$
6. (b) $f(x) = x^2$ if $x \neq 0$ and $f(0) = 2$ on $[0, 2]$
   (d) $f(x) = |x|$ on $[-1, 1]$
14. Find the values of $f(0)$ and $f(1)$, and observe that $f'(x) > 0$
     for all x.

EXERCISES 21 (page 148)

1. (a) $8.1853312 < \sqrt{67} < 8.1853541$
2. First show that $P_4(x) = (x-1) - \frac{1}{2}(x-1)^2 + \frac{1}{3}(x-1)^3 - \frac{1}{4}(x-1)^4$.
    Then $P_4(.5) = -\frac{131}{192}$, $\ln .5 = P_4(.5) + R_4(.5)$, and
    $-\frac{1}{5} < R_4(.5) < -\frac{1}{160}$ imply that $-.882291 < \ln .5 < -.688542$.
4. (a) $R_n = \frac{e^c x^{n+1}}{(n+1)!}$

EXERCISES 22 (page 156)

1. (a) $\frac{1}{3}$
    (b) $\frac{\ln 10}{\ln 2}$
    (d) ∞
2. (c) $\lim_{x \to \infty}(1 - \cos x)$ does not exist
3. (b) $\frac{1}{\sqrt{e}}$

EXERCISES 24 (page 172)

1. $L(f, P) = 3$ for all partitions P while $U(f, P) = 3 + \Delta x_n$
2. $U(f, P_n) = \frac{365}{3} + \frac{225}{2n} + \frac{125}{6n^2}$
3. (a) $L(f, P_n) = \frac{b^3}{6}\left(2 - \frac{3}{n} + \frac{1}{n^2}\right)$
4. $U(f, P_n) = \frac{8}{n^{\frac{3}{2}}}\sum_{i=1}^{n} \sqrt{i}$
8. (d) $\sum_{i=1}^{n} i^4 = \frac{1}{30}n(n+1)(2n+1)(3n^2 + 3n - 1)$.
9. (a) $U(f, P_n) = \sum_{i=1}^{n} \frac{1}{i+n-1}$
    (c) Observe that $\lim_{n \to \infty} \sum_{i=n+1}^{2n} \frac{1}{i} = \lim_{n \to \infty} L(f, P_n) = \int_1^2 \frac{1}{x}\, dx$

EXERCISES 25 (page 180)

6. Observe that $\left| \int_a^b f \right|$ has only two possible values, either $\int_a^b f$
    or $-\int_a^b f$
9. $c = \arcsin \frac{2}{\pi}$

EXERCISES 26 (page 187)

7. Here is part of the proof. For any $x \in T$, $g(x) \le \sup_T g$
    and $\frac{1}{g(x)} \ge \frac{1}{\sup_T g}$. Then $\inf_T \left(\frac{1}{g}\right) \ge \frac{1}{\sup_T g}$ by Lemma 5.2.

EXERCISES 27 (page 191)

4. (c) $\int_{-\pi}^{\pi} \sin^2 nx\, dx = \int_{-\pi}^{\pi} \frac{1 - \cos 2nx}{2}\, dx = \pi$.
5. (a) $F(x) = \frac{1}{2}x^2$ if $0 \le x < 1$, $F(x) = \frac{1}{2}x^2 - x + 1$ if $1 \le x < 2$,
    $F(x) = \frac{1}{2}x^2 - 2x + 3$ if $2 \le x < 3$, etc.

EXERCISES 28 (page 198)

1. (a) Converges to 1
   (b) Diverges to $+\infty$
2. (a) $\int_0^1 e^{-x^2}\, dx$ is proper, and $\int_1^\infty e^{-x^2}\, dx$ converges by comparison with $\int_1^\infty e^{-x}\, dx$.
3. (b) $f(x)$ must have an infinite discontinuity for some value $c \geq 1$.
6. $\Gamma(\frac{5}{2}) = \frac{3}{4}\sqrt{\pi}$
7. Let $t = nx$
9. Let $t = x^3$

EXERCISES 29 (page 207)

1. $\{f_n(x)\}$ converges uniformly to 0 on $[-1, 1]$
3. $\{f_n(x)\}$ converges to $f(x) = 0$ if $x > 0$ and $f(0) = 1$. The convergence is uniform on sets of the form $[b, \infty)$, for $b > 0$.
4. $\{f_n(x)\}$ converges uniformly to 1 on sets of the form $[a, \infty)$ for $a > 1$ and converges uniformly to 0 on sets of the form $[-b, b]$ for $0 < b < 1$. The convergence is not uniform on any set containing $x = 1$.
6. $f(x) = 0$ on $[0, \infty)$. The convergence is uniform on sets of the form $[a, \infty)$ for $a > 0$. There is no uniform convergence on a set containing 0.
7. $\{f_n(x)\}$ converges uniformly to 0 on the set of real numbers.
8. Observe that $\{f_n(x)\}$ has a relative maximum at $(\frac{1}{n}, n)$.

EXERCISES 30 (page 215)

1. (a) Nothing can be said since $f(x)$ is continuous on $[-1, 1]$.
   (b) There can be no uniform convergence on any set containing $+1$ or -1.
2. Consider cases, such as $|x| > 1$, $|x| < 1$, $x = 1$, $x = -1$.
4. The value is 0. Use Theorem 30.3.
5. (a) Use $\int \frac{nx}{1+nx}\, dx = \int \left(1 - \frac{1}{1+nx}\right)\, dx$ and L'Hospital's Rule.
7. Show that $f'(0) \neq g(0)$.

EXERCISES 31 (page 226)

1. $\sum \frac{1}{\sqrt{n}}$
2. (a) $2 < \sum_{n=1}^{\infty} \frac{1}{n^{\frac{3}{2}}} < 3$
   (b) $\sum_{n=1}^{\infty} \frac{1}{n^{\frac{3}{2}}} < 2.7678$

3. Use $\sum_{n=1}^{\infty} \frac{1}{n^{\frac{5}{4}}}$

4. Use the Integral Test.

5. Consider Exercise 4.

6. Only condition (c) guarantees the convergence of $\sum x_n$.

14. $\lim_{n \to \infty} n \left(1 - \frac{x_{n+1}}{x_n}\right) = \frac{3}{2} > 1$.

EXERCISES 32 (page 235)

1. Converges by absolute comparison with $\sum \frac{1}{n^{\frac{3}{2}}}$.

2. $x_{3n} = \frac{9n^2 - 2}{3n(3n-1)(3n-2)}$, and so $\sum x_{3n}$ diverges.

3. Consider the function $f(x) = \frac{\ln x}{x}$.

5. No. Why?

14. More than e^{100}.

EXERCISES 33 (page 240)

1. (b) $\frac{2}{2+x}$ if $|x| < 2$
   (c) $+\infty$

2. Show $\frac{1}{n(n+1)(n+2)} = \frac{1}{2n} - \frac{1}{n+1} + \frac{1}{2(n+2)}$, and then
   $\sum_{n=1}^{k} \frac{1}{n(n+1)(n+2)} = \frac{1}{4} - \frac{1}{2(k+1)} + \frac{1}{2(k+2)}$.

4. Let $S_{2k} = \sum_{n=1}^{2k} \frac{1}{n^3}$ and $T_{2k} = \sum_{n=1}^{2k} \frac{(-1)^{n+1}}{n^3}$.
   Show that $S_{2k} - T_{2k} = \frac{1}{4} S_k$, and hence that $T = \frac{3S}{4}$.

5. Use $S_k + \int_k^{\infty} \frac{1}{x^3} dx$ for upper bounds.

EXERCISES 34 (page 246)

1. (a) Use $M_n = |\sin r|^n$ for $|r| < \frac{\pi}{2}$.
   (b) Observe that $|x \ln x| \leq \frac{1}{e}$ for all $x \in [0, 1]$.
   (c) For $t > 0$, let $M_n = ne^{-nt}$, and prove uniform convergence on $[t, \infty)$.

4. (a) Show that $\int_0^{\frac{\pi}{2}} \frac{\cos nx}{n^2} dx = \frac{(-1)^{n+1}}{(2n-1)^3}$.
   (b) .969 is an approximation with an error less than $\frac{1}{1331} < .001$.

5. Show that $\sum \frac{d}{dx} \left(\frac{1}{n} e^{\frac{x}{n}}\right) = \sum \frac{1}{n^2} e^{\frac{x}{n}}$ converges uniformly on sets of the form $(-\infty, b]$.

6. Hint: $\sum_{n=1}^{\infty} \frac{1}{n^{x^2}}$ is a p-series.

EXERCISES 35 (page 258)

1. (a) $[\frac{1}{2}, \frac{3}{2})$
   (b) $(-\sqrt{3}, \sqrt{3})$
   (c) $(\ln 2, \infty)$
3. $\frac{1}{2} - \frac{1}{72} + \frac{1}{3600} \approx .48583$
4. $\sqrt{x} = 3 + \frac{1}{6}(x - 9) + \sum_{n=2}^{\infty} \frac{(-1)^{n-1} 1 \cdot 3 \cdot 5 \cdots (2n-3)}{2^n 3^{2n-1} n!}(x - 9)^n$
6. $x = -\frac{2}{3}$ yields $\sqrt{3} \approx 1.70976$ using 5 terms, while $x = \frac{1}{3}$ yields $\sqrt{3} \approx 1.731915$ using 5 terms also.
8. (b) $\ln 2$
   (c) $\frac{\pi}{4}$
   (d) $+\infty$
   (h) $\sinh 1$
   (i) $\frac{3}{2}$
   (j) $\sqrt{2}$

EXERCISES 36 (page 268)

2. $A_0 = \frac{1}{2\pi} \int_{-\pi}^{\pi} |x|\, dx = \frac{\pi}{2}$
   $A_n = \frac{1}{\pi} \int_{-\pi}^{\pi} |x| \cos nx\, dx = \frac{2}{\pi} \int_0^\pi x \cos nx\, dx = \frac{2}{\pi n^2}((-1)^n - 1)$
3. Let $x = 0$ in Exercise 2 to find $S = \frac{\pi^2}{8}$.
10. Let $B_n = \frac{1}{\pi} \int_{-\pi}^{\pi} e^x \sin nx\, dx$ and using integration by parts, show that $B_n = \frac{(-1)^{n+1}}{(n^2+1)\pi}(e^\pi - e^{-\pi})$.
14. Begin with $\sin^6 x = \frac{1}{2^6}\binom{6}{3} + \sum_{i=1}^{3} \frac{(-1)^i}{2^5}\binom{6}{3-i}\cos 2ix$.

Index

Abel, N., 23,25, 251
Absolute convergence, 250
Absolute Convergence Test, 230
Absolute value, 2, 3
Accumulation point (see cluster point)
Alternating harmonic series, 72
Alternating series, 228
Apollonius, 21
Archimedean axiom, 46
Archimedes, 21, 117, 157, 159–160, 169
Arithmetic mean sequence, 60
Associative law, 272

Barrow, I., 22, 25
Berkeley, G., 23
Bernoulli, James, 23, 25, 118
Bernoulli, John, 23, 25, 98, 118, 141, 149
Bernoulli number, 18, 20, 265
Biconditional, 274
Binomial theorem, 53
Bolzano, B., 3, 23, 25, 99, 100
Bolzano-Weierstrass Theorem:
     for sets, 45
     for sequences, 58
Borel, E., 3
Bound for a set:
     greatest lower, 40
     least upper, 39
     lower, 40
     upper, 39
Boundary point, 92
Bounded:
     above, 34, 79
     below, 34, 79

function, 79, 108
sequence, 34
set, 40
Brahe, T., 22
Briggs, H., 141

Cantor, G., 24, 25, 260
Cartesian product, 97
Cauchy, A., 21, 23–25, 27, 63, 100, 158, 163–164, 179, 209
Cauchy convergence criterion, 67
Cauchy Mean-Value Theorem, 150
Cauchy sequence, 63
Cavalieri, B., 22, 25, 160–161, 169
Cesaro summable, 60
Chain Rule, 125
Closed set, 93
Closure, 272
Cluster point of a set, 44, 66
Codomain, 97
Combinations, 53
Commensurable, 159
Commutative law, 272
Compact set, 94
Comparison Test:
     for improper integrals, 195–196
     for infinite series, 73, 218–219
Completion axiom, 39
Composite function, 103–104
Conditional, 274
Conditional convergence, 231
Conjunction, 274
Construction theorem, 277
Continuous:
     function, 101
     uniformly, 112

ISBN 0-534-92162-0

90000